For and Against Scientism

Collective Studies in Knowledge and Society

Series Editor: James H. Collier is Associate Professor of Science and Technology in Society at Virginia Tech, USA

This is an interdisciplinary series published in collaboration with the Social Epistemology Review and Reply Collective. It addresses questions arising from understanding knowledge as constituted by, and constitutive of, existing, dynamic, and governable social relations.

Titles in the Series

For and Against Scientism

Science, Methodology, and the Future of Philosophy

Edited by
Moti Mizrahi

ROWMAN & LITTLEFIELD
Lanham • Boulder • New York • London

Published by Rowman & Littlefield
An imprint of The Rowman & Littlefield Publishing Group, Inc.
4501 Forbes Boulevard, Suite 200, Lanham, Maryland 20706
www.rowman.com

86-90 Paul Street, London EC2A 4NE

British Library Cataloguing in Publication Information Available

Library of Congress Cataloging-in-Publication Data on File

ISBN 978-1-5381-6333-7
ISBN 978-1-5381-6334-4 (electronic)

Contents

Chapter 1

Introduction

Moti Mizrahi

The term *scientism* is used in several ways. It is used to denote an episte-mological thesis according to which science is the source of our knowledge (or some other epistemic good, such as justified belief) about the world and ourselves. Relatedly, it is used to denote a methodological thesis according to which the methods of science are superior to the methods of non-scientific fields or areas of inquiry. It is also used to denote a metaphysical thesis according to which what exists is what science says exists. In recent decades, the term *scientism* has acquired a derogatory meaning when it is used in defense of non-scientific ways of knowing. In particular, some theologians and philosophers level the charge of "scientism" against those (mostly scientists) who are dismissive of religion and philosophy. Other philosophers, however, embrace scientism, or some variant thereof, and object to the pejorative use of the term *scientism*.

1. *SCIENTISM* AS A PEJORATIVE TERM

According to Robert Pennock (1996, 551), the term *scientism* is

> A term of derision coined by Hermeneutic critics of science to label those who wanted to apply the methods of the natural sciences 'inappropriately' to the human sciences, for which they thought the literary model of Hermeneutic *interpretation* should reign as the proper method. (emphasis in original)

Attempts to apply the empirical methods of the sciences to the human sciences are probably as old as modern science itself. For example, inspired by the work of Isaac Newton, David Hume attempted "to introduce the

1

experimental method of reasoning into moral subjects," as the full title of his *Treatise on Human Nature* (1739–1740) clearly states. Additional examples include Baruch Spinoza's application of the geometrical method in the *Ethics* (1677) and Duncan Macdougall's experimental attempt to weigh the human soul (special to *The New York Times*, 1907).

When Robert Pennock (1996, 551) says that the term *scientism* is "a term of derision," his target of criticism is Phillip Johnson whose book, *Darwin on Trial* (1991), is the inspiration for the brand of creationism known as "Intelligent Design." But there are earlier uses of the term *scientism* as a derogatory term by people of faith. For instance, in their book *Roadblocks to Faith* (1954), James A. Pike and John McGill Krumm distinguish science from scientism and claim that the latter is a religion. Around the same time, H. Richard Rasmusson (1954, 393) characterizes scientism as "a cult that has made a religion out of science." Responding to Rasmusson's article, Addison Gulick (1955, 392) observes that the "word *scientism* seems to be intended approximately as a disparaging nickname for evolutionary naturalism."

Those who use the term *scientism* as a derogatory term, then, are concerned about science encroaching on fields or areas of inquiry that presuppose the existence of the very things whose existence they take science to be questioning or denying, such as God, the supernatural, and the like. As W. H. Werkmeister (1959, 21) puts it, with the help of a quote from Russell Davenport's *The Dignity of Man* (1955):

> Here there has emerged a scientific reductionism which falsifies the picture of man and of his position in the universe; this "scientism" implies that "the great ideals for which humanity has fought and suffered—the Good, the Beautiful, and the True—honor, love, and sacrifice—the eightfold path of Buddha, the Christian vision of redemption, the Moslem communion with God—all these inspired revelations and ideals are just illusions on the surface of reality." (Davenport 1955, 176)

In that respect, as far as the scientism debate is concerned, the term *science* typically covers the physical sciences, such as physics and chemistry; the life sciences, such as biology and genetics; the social sciences, such as psychology and sociology; to the exclusion of the arts and the humanities, such as literature, philosophy, and religious studies.

More recently, one can find theologians and scholars of religion writing in the pages of philosophy of religion journals, such as *Zygon* and *International Journal for Philosophy of Religion*, that making science the enemy of religion has turned out to be a failed strategy. Instead, they argue that the enemy of religion is scientism, by which they mean naturalism, materialism, and/or secularism (e.g., see Barbour 2001; Haught 2005; Burch 2016).

A few decades later, from a term deployed when defending religion from science intruding on its territory, the term *scientism* became a term used in defense of philosophy against science as well. For instance, according to Tom Sorell (1991, x), scientism "is a matter of putting too high a value on science in comparison with other branches of learning or culture," such as philosophy. Understood in this way, scientism is neither a thesis nor a stance but rather an attitude (Peels 2017). The term *scientism* is then used as a pejorative term that is supposed to pick out an attitude that is condemned as excessive or exaggerated *by definition*. As Susan Haack (2007, 17–18) puts it, "Scientism is an exaggerated kind of deference towards science, an excessive readiness to accept as authoritative any claim made by the sciences, and to dismiss every kind of criticism of science or its practitioners as anti-scientific prejudice" (see also Hua 1995, 15). Philosophers who level the charge of "scientism" against those who deride philosophy typically identify prominent scientists, such as Stephen Hawking and Neil deGrasse Tyson, as exhibiting this kind of attitude toward science (Kidd 2016).

However, as René van Woudenberg et al. (2018, 2) rightly point out,

> No one will accept this notion of "scientism" as an adequate characterization of their own views, as no one will think that their deference to science is *exaggerated*, or their readiness to accept claims made by the sciences is *excessive*. (emphasis in original)

In fact, Susan Haack (2012, 76) herself observes that, before it was weaponized by those who sought to defend religion and philosophy from science trespassing on their territories, "the word 'scientism' was neutral." And so, some contemporary philosophers have argued that the term *scientism* should remain neutral. For instance, Rik Peels (2017, 11) argues that we should treat scientism as a neutral thesis, rather than an attitude, since whether it is rational, warranted, or permissible to have such an attitude is a question that should be "up for debate." Moreover, Peels (2018, 29) argues, "No matter how one understands 'scientism' [that is, as an attitude, an affection, or a stance], it will always imply some scientistic thesis or other [for example, that 'we should have a scientistic attitude']." Likewise, Moti Mizrahi (2017a, 352) argues that to define the term *scientism* in a pejorative way is to provide a persuasive definition of *scientism*, to beg the question against anyone who is inclined to endorse scientism, and to turn the scientism debate into a mere verbal dispute.

For these reasons, and because "the truth or falsity of scientism matters a great deal" (van Woudenberg et al. 2018, 1), this book follows Mikael Stenmark (1997), Rik Peels (2017; 2018), and Moti Mizrahi (2017a) in treating scientism as a thesis that is up for discussion and debate rather than an attitude that is rendered mistaken *by definition*. Indeed, some have used the term *scientism* neutrally. For example, the historian Richard Olson (2008, 1) uses the term *scientism* to refer to "the transfer of ideas, practices, attitudes, and methodologies from the context of the study of the natural world [. . .] into the study of humans and the social institutions, without imposing any judgment on the legitimacy of such as appropriation." It is part of the philosophical debate over scientism to determine whether such an appropriation is in fact legitimate. Taken as a thesis that is up for debate, then, scientism is typically understood as an epistemological thesis, a methodological thesis, or a metaphysical thesis. This chapter will briefly introduce readers to these three scientistic theses in order.[1]

This chapter will be devoted to an introductory discussion of the scientism debate in philosophy. It is important to note, however, that discussions of scientism have occurred, and still occur, beyond philosophy. For example, in his *Social Statics; or, the Conditions Essential to Human Happiness Specified, and the First of Them Developed* (first published in 1850), the sociologist Herbert Spencer develops a sort of scientism insofar as, for him, what is true of the forces acting in nature "is equally true of the agencies acting in society" (1873, 426). Like Thomas Hobbes before him, who sought to apply the concepts of the new science of the seventeenth century, such as *conatus* (or endeavor), to social and political philosophy (Bernstein 1980), Spencer sought to apply the scientific concept of energy to society. According to Harold I. Sharlin (1976, 457), Spencer was a "spokesman" for the "new scientism [of the nineteenth century]" because he advocated "the application of the principles derived from the natural sciences to other disciplines." Another figure that might be regarded as a spokesperson for the new scientism of the nineteenth century, and that should be mentioned, is Auguste Comte. Comte shared with Spencer the idea that science can help solve social and political problems, as the title of his *Plan for the Scientific Work Necessary to Reorganize Society* (written in 1822)—also called *First System of Positive Polity*, from which the label *positivism* derives—suggests. The idea that science has an important role to play in reshaping society was later taken up by a group of philosophers known as logical positivists.

Outside of philosophy, there are also those who do not advocate but rather decry the encroachment of science into domains traditionally perceived as non-scientific. For example, in his "Scientism and the Study of Society" (Part I published in 1942), the economist F. A. von Hayek laments what he calls "that slavish imitation of the method and language of Science" (1942, 269). Like the aforementioned religious critics, Hayek is careful to distinguish between science and scientism. He does not object to "the methods of Science in their proper sphere" (Hayek 1942, 269). Rather, he is concerned with what he calls "the scientistic prejudice," which, "before it has considered its subject, claims to know what is the most appropriate way of investigating it" (Hayek 1942, 269).[2]

2. *SCIENTISM* AS AN EPISTEMOLOGICAL THESIS

First occurrences of the term *scientism* suggest that it was used to denote an epistemological thesis, in particular, a thesis concerning how knowledge (or some other epistemic good, such as justified belief) about the world and ourselves is acquired. For instance, in Stephen Pearl Andrews' *The Primary Synopsis of Universology and Alwato: The New Scientific Universal Language* (1871, xiii), *scientism* is defined as "the Spirit or Principle of Science—regular, exact, precise, etc." By *science*, Andrews (1871, 19) means "a Systematic, Orderly, and somewhat Complete Arrangement of what is certainly known or held to be known, and of what is important to be known, in respect to the particular subject or Department of Being treated of."[3] So, the essence of science, according to Andrews, is regular, exact, precise (or systematic) knowledge. Around the same time, Henry N. Day (1870, 513–514) uses the term *scientism* to refer to an epistemological view that gives priority to "observation over reflection." More recently, and along the same lines, David Bell (1962, 50) defines *scientism* as "the belief that reason must subordinate itself to 'experience' conceived as a bloodless and unrealistic succession of sense experiences or observations giving rise to empirical generalisations."

Construed as an epistemological thesis, then, scientism is a species of empiricism.[4] More specifically, it is the view that gives priority to scientific forms of knowledge and ways of knowing—such as observation and experimentation—over non-scientific forms of knowledge and ways of knowing—such as armchair reflection. As Patricia Churchland (2011, 4) puts it, "philosophy and science are working the same ground, and evidence should trump armchair reflection." As an epistemological thesis, however, scientism is a narrower thesis than empiricism. This is because empiricism allows non-scientific but empirical modes of inquiry to count as knowledge-producing, whereas scientism grants knowledge-producing status to scientific modes of inquiry only. To illustrate this point, Moti Mizrahi (2017a, 354) gives the following example with respect to observation:

> Compare, for instance, the sort of telescopic observations Galileo conducted when he discovered the rings of Saturn with simply lying around and gazing at the night sky. It might seem as if the former is "scientific," whereas the latter is not, but that such gazing can still produce "real knowledge."

Accordingly, observation by means of scientific instruments, such as telescopes, is a scientific mode of observation, and hence knowledge-producing on epistemological scientism.

Understood as an epistemological thesis, scientism can vary in terms of how superior to non-scientific ways of knowing science is taken to be. In his review of Tom Sorell's *Scientism: Philosophy and the Infatuation with Science* (1991), Philip Kitcher (1991) relates the following story that, for him, captures what is at stake in the debate about the limits (or lack thereof) and scope of science:

> About 20 years ago, in an informal seminar held in connection with a drama festival at the University of Cambridge, one of the most prominent literary critics of our time suggested that Moliere and Stendhal have more to teach us about the workings of the human mind than any conglomeration of academic psychologists. (Kitcher 1991, 118)

Accordingly, to say that *only* psychology can teach us about the workings of the human mind, whereas Moliere and Stendhal can teach us nothing about that, is to subscribe to a strong version of *epistemological scientism*. According to this view, "Science is the *only* source of justified belief or knowledge about ourselves and the world" (de Ridder 2014, 25; emphasis added). In other words, "only certifiably scientific knowledge counts as real knowledge" (Williams 2015, 6). As Rik Peels (2018, 34) points out, other variants of *epistemological scientism* include the following:

(a) All genuine knowledge is to be found only through (methods of) the natural sciences. (See also Stenmark 1997, 19.)
(b) The natural sciences provide the only reliable path to knowledge. (See also Rosenberg 2011, 6.)
(c) All questions can in principle be answered by the natural sciences. (See also Atkins 1995.)
(d) Everything that can be known can be known through the natural sciences. (See also Russell 1946, 863.)

Along the same lines, Mikael Stenmark (1997) labels the "view that the only reality we can know anything about is the one science has access to" *epistemic scientism* (19) and the view that "we are rationally entitled to believe only what can be scientifically proved or what is scientifically knowable" *rationalistic scientism* (21).

As Rik Peels (2018) also points out, each of these statements of epistemological scientism can take a weak or a strong version as well. Take (b), for example. The claim that science provides the *only* reliable path to knowledge is stronger than the claim that science provides the *best* reliable path to knowledge. Unlike the former, which rules out non-scientific paths to knowledge, the latter allows for paths to knowledge other than science, but it says that, of all the paths to knowledge, science is the best one. Similarly, Moti Mizrahi (2017a) distinguishes between *Strong Scientism*, which is the view that scientific knowledge is the *only* knowledge we have, and *Weak Scientism*, which is the view that scientific knowledge is the *best* knowledge we have. The former implies that those things we call knowledge that are not scientific are not really knowledge at all, whereas the latter implies that there is non-scientific knowledge, although scientific knowledge is better than non-scientific knowledge. A similar difference holds between the claim that science "is much the most valuable part of human learning" (Sorell 1991, 1) and the claim that "science is the *only* valuable part of human learning" (Sorell 1991, 1; emphasis in original), although Tom Sorell (1991) does not always treat them differently. The latter implies that non-scientific ways of learning have no value at all, whereas the former implies that non-scientific ways of learning are valuable but not as valuable as scientific ways of learning about the world and ourselves. Contemporary philosophers in the analytic tradition who subscribe to a strong version of epistemological scientism include James Ladyman et al. (2007, 61), "who admire science to the point of frank scientism," and Alexander Rosenberg (2011; 2017).

In addition to the strong/weak distinction, some philosophers distinguish between versions of scientism as an epistemological thesis in terms of the scope of the thesis. For example, Stenmark (1997) distinguishes between "academic-internal" and "academic-external" versions of epistemological scientism. The former is the view that "all, or at least some, of the genuine, non-scientific academic disciplines can eventually be reduced to (or translated into) science proper" (Stenmark 1997, 17), whereas the latter is the view that "all or, at least, some of the essential non-academic areas of human life can be reduced to (or translated into) science" (Stenmark 1997, 18). Along similar lines, Peels (2018) draws a distinction between "academic" and "universal" versions of epistemological scientism. The former "is restricted to the academic disciplines," whereas the latter "is meant to apply both inside and outside of the academy" (Peels 2018, 31). For example, to claim that metaphysics should be turned into physics, as Otto Neurath (1987) does, is to advocate an academic (or academic-internal) version of epistemological scientism. Contemporary philosophers in the analytic tradition who subscribe to an academic version of epistemological scientism include Wesley Buckwalter and John Turri (2018, 281). Buckwalter and Turri (2018, 281) defend a version of scientism they call "moderate scientism," which is the view that "science can help answer questions in disciplines typically thought to fall outside of science," such as academic philosophy.

3. *SCIENTISM* AS A METHODOLOGICAL THESIS

In addition to *scientism* as an epistemological thesis, early instances of the term *scientism* suggest that it was used to denote a related methodological thesis. For instance, in his definition of *scientism* in *The Primary Synopsis of*

Universology and Alwato: The New Scientific Universal Language, Stephen
Pearl Andrews (1871, xiii) describes the methods of science as "regular,
exact, precise, etc." Accordingly, scientism as a methodological thesis captures the fact that the scientism debate is partly about "the idea that science,
or the scientific method, is superior to all other modes of inquiry" (Beale
2017, 67). That is to say, if the methods and practices of science are indeed
more "regular, exact, [and] precise" (Andrews 1871, xiii) than non-scientific
modes of inquiry, then the former can be said to be superior to the latter in
those respects.

Another indication that early instances of the term *scientism* were used
to denote a methodological thesis about scientific methods and practices is
the following definition of "scientism" from *The Century Dictionary: An
Encyclopedic Lexicon of the English Language* (edited by William Dwight
Whitney, 1890): "The views, tendency, or practice of scientists [Recent.]"
(Vol. V, 5397). The following example is then provided: "Mr. Harrison's
earnest and eloquent plea against . . . the exclusive *scientism* which, because
it cannot find certain entities along its line of investigation, asserts loudly that
they are either non-existent or 'unknowable', is strong." This example hints at
both the empiricist element of scientism discussed in section 2 above, which
emphasizes observation and experimentation over armchair reflection, and
the fact that there were early critics of scientism who resisted the expansion
of scientific practices and methods of investigation into other domains.

According to both Mikael Stenmark (1997) and Rik Peels (2018), *methodological scientism* entails a demand that non-scientific disciplines adopt
the methods and practices of science. As Stenmark (1997, 17) puts it, methodological scientism (or "academic-internal scientism") is the "attempt to
extend the use of the methods of natural science to other academic disciplines." Likewise, according to Peels (2018, 31), "the methodological variety
[of scientism] grants that [non-scientific fields] are proper academic disciplines that ask sensible questions, [but] it asserts that they are so only if they
adopt the methods of the natural sciences, such as observation and experimentation." On the other hand, Moti Mizrahi (2017a; 2017b; 2018a; 2018b;
2018c) argues that non-scientific disciplines would benefit from adopting the
methods and practices of science, but it is not a necessary condition for the
production of knowledge in those disciplines. On Mizrahi's *Weak Scientism*,
non-scientific disciplines are still proper academic disciplines that produce
knowledge, even if they do not use the methods and practices of science to
do so, it's just that the knowledge produced by scientific disciplines is better
(both quantitatively and qualitatively) than the knowledge produced by non-
scientific disciplines.

Opponents of scientism often point out that the methods of science have
limits. As Patricia Churchland (2011, 3) puts it, "Scientism, as I have been

duly wagged, is overreaching." That is to say, there are certain questions that are beyond the scope of scientific inquiry, such as questions about the existence of God and the supernatural (Stenmark 2016, 1–17). As Susan Haack (2012) puts it:

> Environmental science can't, by itself, tell us whether the benefits of damming the river outweigh the drawbacks, and certainly not whether building the dam is a good idea; medical science can't, by itself, tell us whether abortion is morally acceptable (nor, of course, whether it should be legally permitted); economics can't, by itself, tell us whether we should change the tax system in this way or that. (Haack 2012, 90)

In "Science Unlimited?" Carl G. Hempel (1973, 35) argues that this point "rests on a too simple-minded view of scientific testing." Even the "hypothesis that there are ghosts who do appear, but only in the absence of scientists," Hempel (1973, 35) argues, can be tested indirectly "by means of suitable automatic recording instruments." Of course, the ghost hypothesis can be made untestable by stipulating that these ghosts cannot be detected by any observer or instrument whatsoever. However, that is tantamount to making the ghost hypothesis not only untestable but also unknowable, Hempel (1973, 35) argues, for then there would be no grounds on which such a claim to knowledge can be based, including "logical or conceptual grounds." In general, Hempel (1973, 34) argues, insofar as the methods of science are supposed to be *ampliative*, that is, they are supposed to take us from what is known to what is unknown, they cannot be said to be limited without thereby being committed to the unwanted consequence that nothing is knowable.

Like epistemological scientism, methodological scientism can take a weak or a strong version as well. According to the *strong* variant of methodological scientism, *only* scientific methods yield knowledge (or some other epistemic good, such as justified belief) about the world and ourselves. This strong version implies that, if non-scientific fields or areas of inquiry are to produce knowledge about the world and ourselves, they must use scientific methods. In that respect, the strong variant of methodological scientism raises concerns about "scientific imperialism" (Kitcher 2012) or "scientific expansionism" (Stenmark 2004, xi–xii), which is when science goes beyond what is considered to be its proper sphere. For this reason, some philosophers have deemed it necessary to defend the importance of the humanities and the arts in education. For example, in her *Not For Profit: Why Democracy Needs the Humanities*, Martha Nussbaum (2010, 7) tries "to show how the humanities and arts are crucial both in primary/secondary and in university education." She is careful to point out that she does not "deny that science and social science [. . .] are also crucial to the education of citizens" (Nussbaum 2010, 7).

However, unlike the humanities and arts, Nussbaum observes, science and the social sciences are not in a state of crisis.

According to the *weak* version of methodological scientism, scientific methods are *better* than non-scientific methods at producing knowledge (or some other epistemic good, such as justified belief) about the world and ourselves. An advocate of weak methodological scientism, then, can grant that "there is plenty of good, solid work in non-scientific disciplines such as history, legal scholarship, music theory, etc." (Haack 2012, 79), but insist that such work can be improved by the introduction and application of scientific methods (Mizrahi 2018a; 2018b). For instance, Wesley Buckwalter and John Turri (2018) argue that, as far as academic philosophy is concerned, what Mikael Stenmark (2004, xi–xii) calls "scientific expansionism," that is, that "the boundaries of science can and should be expanded in such a way that something that has not been understood as science can now become a part of science" (cf. Peels 2018, 28–29), has proved to be quite beneficial to the field. That is to say, according to Buckwalter and Turri (2018, 292), "Scientific research has promoted significant progress in philosophy and its further development within the field should be welcomed and encouraged."

As in the case of epistemological scientism, methodological scientism can also vary along the internal/external or academic/universal dimensions. To endorse internal (or academic) methodological scientism is to advocate for the application of scientific methods and practices to academic disciplines that are traditionally understood as non-scientific (Stenmark 1997, 17). To endorse external (or universal) methodological scientism is to advocate for the application of scientific methods and practices to all areas of inquiry, academic and otherwise (Peels 2018, 31). In that respect, insofar as they advocate for the application of experimental methods to philosophy, Buckwalter and Turri's (2018, 281) "moderate scientism" is an instance of internal (or academic) methodological scientism.

4. *SCIENTISM* AS A METAPHYSICAL THESIS

In addition to *scientism* as epistemological and methodological theses, early instances of the term *scientism* suggest that it was used to denote a metaphysical thesis as well. For instance, in *The Science of Aesthetics* (1872, 425), Henry N. Day uses the term *scientism* to refer to the thesis that "mind and matter are made one and identical," which Day finds objectionable on the grounds that it ignores the distinction between substance and form, and thus renders the human person "a mysterious inexplicable organism which is neither mind nor matter." Another indication that early instances of the term *scientism* were used to denote a metaphysical thesis about the nature of reality is a passage from *Morning Light: A New-Church Weekly Journal* from 1878 (Vol. I) in which *scientism* is described as "the material philosophy [that] threatens to dominate in every sphere of human thought" (134). More recently, John Dupré (1988) aligns *scientism* with *physicalism*, which is a thesis about "the physical nature of the mental" (32), and *materialism*, which is the thesis that, "if one removed from the universe all the physical entities, there would be nothing left" (33).

As René van Woudenberg et al. (2018, 10) observe, scientism "has a history." Understood as a metaphysical thesis, the history of scientism can be traced back to those philosophers, such as Thomas Hobbes, which Catherine Wilson labels "old materialists." As Wilson (2017, 113) writes, the "old materialists refused to see misery and oppression as consistent with divine justice and a Providential plan rather than as reflecting the various powers and choices of its human inhabitants." Materialism, Wilson (2017, 113) argues, was "the philosophy of what might be called the 'party of humanity'," since it sought to dismantle, and thereby liberate human beings, from idealized hierarchies and dividing lines. Despite being the "party of humanity," Wilson (2017, 113) argues, materialism has been seen as a "reductive, demeaning, pessimistic philosophy," and so "the fear of reductionism and scientism remains alive" as well as "the perceived threat of 'scientism' in the humanities and in the general culture" (Wilson 2017, 115).

Construed as a metaphysical thesis, then, *scientism* is a species of materialism, physicalism, or naturalism more broadly,[5] according to which "reality is exhausted by the natural world, in the sense of the world as the natural sciences are capable of revealing it to us" (McDowell 1998, 175). This sort of naturalism that restricts reality to what science can discover is what John McDowell casually calls "philistine scientism." According to McDowell (1998, 72):

> A scientific conception of reality is eminently open to dispute. When we ask the metaphysical question whether reality is what science can find out about, we cannot, without begging the question, restrict the materials for an answer to those that science can countenance.[6]

Contrary to McDowell, James Ladyman et al. (2007, 30) argue that "Naturalism requires that, since scientific institutions are the instruments by which we investigate objective reality, their outputs should motivate all claims about this reality, including metaphysical ones." In other words, science, especially physics, is our guide to what exists. As Alyssa Ney (2018, 263) puts it, "physics is the final arbiter of what is true and real."[7] Although Ney (2018) wants to distinguish between this claim and the weaker version of *physicalism* according to which the physicalist "can acknowledge with modesty the possibility of gaining valuable information from other domains of inquiry" (Ney 2018, 278). In that respect, Ney's view

Table 1.1   Varieties of Scientism

	Scientism		
	Epistemological	*Methodological*	*Metaphysical*
Strong	Scientific knowledge is the only knowledge we have.	Scientific methods are the only methods that produce knowledge (or some other epistemic good, such as justified belief).	Science is the only guide to the nature of reality.
Weak	Scientific knowledge is better than non-scientific knowledge.	Scientific methods are better than non-scientific methods in producing knowledge (or some other epistemic good, such as justified belief).	Science is the best guide to the nature of reality.

amounts to a weak version of metaphysical scientism. According to this weaker variant of metaphysical scientism, reality is not exhausted by our scientific ontology (e.g., the mental may be real, albeit not fundamental), but science is still "our *best* guide to what there is" (Maddy 2005, 444).

As in the case of epistemological and methodological scientism, metaphysical scientism can also vary along the internal/external or academic/universal dimensions. To endorse internal (or academic) metaphysical scientism is to argue that science is the measure of all academic things. To endorse external (or universal) metaphysical scientism is to argue that science is the measure of all things, academic and otherwise. According to the latter, science can tell us not only whether things such as quarks and viruses exist but also whether things such as free will and values exist (Peels 2018, 36).

Some contemporary philosophers in the analytic tradition endorse both epistemological scientism and metaphysical scientism. According to Alex Rosenberg (2017, 213), for example, scientism is the combination of naturalism and empiricism. Likewise, for James Ladyman (2011, 97), scientism "incorporates elements of both empiricism and materialism." Together with Don Ross, David Spurrett, and John Collier, Ladyman adopts a "scientistic stance," which is "the synthesis of the empiricist and materialist stances" (2007, 99). Needless to say, physicalism, naturalism, and materialism, as well as empiricism, are by no means generally accepted among contemporary academic philosophers in the analytic tradition (Bourget and Chalmers 2014).[8] Table 1.1 summarizes the varieties of scientism discussed above.

5. KEY ARGUMENTS FOR AND AGAINST SCIENTISM

Arguments for and against scientism usually trade on what proponents take to be the successes of science and what opponents take to be the internal inconsistency or the negative effects of endorsing scientism. Proponents of scientism argue that the success of science is evidence for scientism. For example, according to James Ladyman et al. (2007):

> Powerful explanations and successful predictions have been produced by sciences that aren't physics and which refer to such entities. They are good ammunition for an epistemic success argument in favour of naturalism [that is, scientism as a metaphysical thesis]. (Ladyman et al. 2007, 41)

In other words, the fact that our best scientific theories successfully explain and accurately predict natural phenomena is evidence that the methods and practices of science are superior to non-scientific modes of inquiry and that the natural world is pretty much the way science says it is, given that other fields or areas of inquiry, such as metaphysics, "can claim no such success" (Ladyman et al. 2007, 7).

René van Woudenberg et al. (2018, 17) criticize this argument for scientism from the success of science by claiming that "it hardly needs saying that a claim about the comparative success of science nowhere near implies that science is our *only* route to knowledge, rational belief, etc." (emphasis added). It is true that the success of science does not *entail* that science is the only way of knowing about the world and ourselves, but it is not clear that this is what the argument from the success of science amounts to. That is to say, the argument for scientism from the success of science need not be a deductive argument whose premises entail its conclusion. Rather, the success of science is supposed to be strong evidence for (or a good reason to accept) scientism. Similarly, the conclusion of the argument for scientism from the success of science need not be construed as *Strong Scientism*, namely, the thesis that science is the *only* way of knowing about the world and ourselves. Rather, it can be construed as *Weak Scientism*, namely, the thesis that science is the *best* way of knowing about the world and ourselves. Understood as a non-deductive argument for *Weak Scientism*, the argument from the success of science might provide strong reasons, namely, the explanatory, predictive, and instrumental success of science, to accept *Weak Scientism* (Mizrahi 2017a; 2017b; 2018a; 2018b; 2018c).

Peter Harrison (2018) also criticizes the argument from the success of science. Harrison (2018, 13) argues that the "history of science is not the place to turn for those seeking an evidentiary basis for ontological naturalism [or metaphysical scientism]." This is because "historical actors did not recognize the natural-supernatural distinction in the way we presently do," Harrison (2018, 8) argues, and because, throughout the history of science, supernatural explanations were often given to natural phenomena (e.g., the role of God in Newtonian mechanics).

By contrast, Moti Mizrahi (2017a; 2017b; 2018a; 2018b; 2018c) argues that, in addition to the explanatory, predictive, and instrumental success of science, there is another sense in which scientific disciplines can be said to be more successful or better than non-scientific disciplines. As Mizrahi observes, in general, scientific disciplines are *quantitatively* better than non-scientific disciplines in terms of research output and research impact. That is to say, scientific disciplines produce more research overall (as measured by publications) than non-scientific disciplines do and the research they produce has more impact overall (as measured by citations) than non-scientific research has. Together with the qualitative superiority of scientific knowledge over non-scientific knowledge (in terms of explanatory, predictive, and instrumental success), Mizrahi argues, the quantitative superiority of scientific knowledge over non-scientific knowledge provides good reasons for accepting *Weak Scientism.*

To the success of science, Rik Peels (2017, 14) adds the following reasons for endorsing scientism:

1. Science is highly successful.
2. The applications of science are everywhere.
3. Beliefs based on science can be tested or corroborated.
4. Many scientific results are counterintuitive.
5. Science has safety mechanisms.
6. We understand the genesis of scientific knowledge.
7. Common sense beliefs display vast disagreement.
8. Science provides debunking explanations of commonsense beliefs.
9. Science shows common sense to be permeated with biases.
10. Science demonstrates that many commonsense beliefs are illusory.

Peels (2017, 20) argues that reasons (8), (9), and (10) in particular could provide good reasons to endorse either a weak or a strong version of (epistemological, methodological, or metaphysical) scientism.

A common argument against scientism is that it is internally inconsistent. For instance, Jeroen de Ridder (2014, 27) argues that "scientism suffers from self-referential problems." That is:

> Not being a scientific claim itself, it would seem scientism cannot be known by anyone. This raises the question of why anyone should assert or believe it in the first place. (de Ridder 2014, 27)

Contrary to de Ridder, however, Moti Mizrahi argues that there can be scientific evidence for scientism and that there is nothing self-referentially incoherent or inconsistent about that. First, one could argue that scientific knowledge is better than non-scientific knowledge on the basis of research output and research impact. Insofar as such an argument would be an inductive argument from samples of scientific research compared to non-scientific research, it would be a scientific argument for scientism (Mizrahi 2017a, 355–362). Second, if there can be no scientific evidence for scientism, since that would be self-referentially incoherent, then there can be no proof for rules of inference in logic, either. As Stathis Psillos (1999, 82) argues, following R. B. Braithwaite (1953), premise-circular arguments, in which "one claims to offer an argument for the truth of α, but explicitly *presupposes* α in one's premises," are viciously circular, whereas rule-curricular arguments, in which "the argument itself is an instance of, or involves essentially an application of, the rule of inference vindicated by the conclusion," are not viciously circular (Psillos 1999, 82). This distinction between premise-circularity and rule-circularity makes it legitimate to use logic to prove rules of inference in logic, given that even "deductive inference is only defensible by appeal to deductive inference" (Ladyman 2002, 49). Therefore, if it is legitimate to use logic to prove rules of inference in logic, then it should also be legitimate to use scientific methods to support a meta-scientific thesis, such as scientism, for both involve a kind of rule-circularity that is not vicious (Mizrahi 2017a, 362–363). Third, weaker versions of scientism, according to which scientific ways of knowing are *better* than non-scientific ways of knowing, are immune from the self-referential incoherence charge. For they allow that there are ways of knowing other than scientific ones, and that there is knowledge other than scientific knowledge, but they insist that scientific ways of knowing are better than non-scientific ways of knowing, and that scientific knowledge is better than non-scientific knowledge (Mizrahi 2018b, 41–42). Finally, to insist that scientism must be supported by non-scientific evidence is to presuppose some sort of foundationalism about knowledge and justification. Proponents of scientism, however, need not accept that kind of foundationalism about knowledge and justification. They could appeal to coherentism or reliabilism to argue that our scientific knowledge coheres with our non-scientific knowledge or that our scientific ways of knowing are generally reliable.

In that respect, critics of scientism often point out that science presupposes that our cognitive faculties, including perception and reasoning, are reliable. Critics of scientism take this as a reason to think that not all of our knowledge is scientific, and hence that epistemological scientism is false. As Mary Midgley (1992) puts it:

> Science cannot stand alone. We cannot believe its propositions without first believing in a great many other startling things, such as the existence of the external world, the reliability of our senses, memory and informants, and the

validity of logic. If we do believe in these things, we already have a world far
wider than that of science. (Midgley 1992, 108)

Of course, the same can be said about any form of human inquiry, both scientific and otherwise. Given that the inquirers are human beings, both scientific
inquiry and non-scientific inquiry rely on human cognitive faculties. Again,
proponents of scientism, especially *Weak Scientism*, according to which
scientific ways of knowing are better (or more reliable) than non-scientific
ways of knowing about the world and ourselves, need not accept this sort
of foundationalism about knowledge and justification. They could appeal
to coherentism or reliabilism to argue that our scientific knowledge coheres
with our non-scientific knowledge or that our scientific modes of inquiry are
generally more reliable than our non-scientific modes of inquiry.

When it is not taken to be a risk in itself (e.g., see Beltramini 2019),
arguments against scientism often focus on what critics perceive to be the
negative consequences of endorsing scientism. For example, Philip Kitcher
(2012) claims that scientism, which he distinguishes from science but does
not define, makes us underestimate "the impact of the humanities and the
arts" and "inspires scientific imperialism." According to John Dupré (2001,
16), scientific imperialism is "the tendency for a successful scientific idea
to be applied far beyond its original home." Likewise, Renia Gasparatou
(2017) argues that scientism, by which she means the "excessive admiration
of science that is based on a naive, exaggerated conception of it as a single,
non-speculative practice or method, which can and will answer all possible
questions" (Gasparatou 2017, 800), prevents students from learning "the skill
of data-theory coordination" (Gasparatou 2017, 802).

Two important points about these criticisms of scientism should be made.
First, more often than not, the critics use the term *scientism* in its derogatory
sense (e.g., see Gasparatou 2017). As René van Woudenberg et al. (2018, 2)
rightly point out, however, "no one will accept this notion of 'scientism' as an
adequate characterization of their own views, as no one will think that their
deference to science is *exaggerated*, or their readiness to accept claims made
by the sciences is *excessive*" (emphasis in original). Second, even if it is true
that accepting scientism will have negative consequences, it is not clear how
that is supposed to be evidence relevant to whether scientism is true or not.
Merrilee H. Salmon (2013, 209) calls this sort of argument the "fallacy of
mistaking some consequence of a belief with evidence for it." We commit this
fallacy when we accept or reject a claim "because of its consequences—the
harm or good that might be caused by holding the belief" (Salmon 2013, 209).

Nevertheless, as mentioned above (see section 1), some philosophers see
scientism as a threat (e.g., see Kitcher 2017). Given that critics of scientism
see scientism as a threat to the humanities and the arts, the question is whether

the fear of scientism is well-founded. As we have seen in section 4, in its metaphysical guise, scientism is a close cousin of materialism. And materialism, according to Catherine Wilson (2017), is the philosophy of "the party of humanity." That is, contrary to the reputation of materialism as a reductive and pessimistic philosophy, Wilson (2017, 113) argues, the "old materialists," such as Thomas Hobbes and Baruch Spinoza, "refused to see misery and oppression as consistent with divine justice and a Providential plan rather than as reflecting the various powers and choices of its human inhabitants." In other words, far from being oppressive, Wilson (2017) argues, materialism is a liberating philosophy, freeing human beings from idealized hierarchies, such as the Great Chain of Being. Another positive consequence of scientism as materialism might be "relief from the 'gendered devaluation of nature' [(Willey 2017, 146)] of traditional philosophy and religion" (Wilson 2017, 125). That is, to insist that "nature and culture are [. . .] the appropriate disciplinary objects of the natural sciences and humanities respectively" is to reinforce "the nature/culture binary" (Willey 2017, 146). On the other hand, to understand "previously disciplinary phenomena as complex 'intra-actions' and 'the world' as an ongoing achievement of (inextricably) naturecultural becomings" is to usher in a genuinely interdisciplinary, "naturecultural" research and scholarship (Willey 2017, 146).

For critics of scientism, however, it is an instance of "scientific imperialism" (Kitcher 2012) or "scientific expansionism" (Stenmark 2004, xi–xii) when science encroaches on non-scientific territories. For proponents of scientism, on the other hand, the introduction of scientific methods into non-scientific fields or areas of inquiry is seen as positive, not negative. Proponents of scientism argue that, if we want non-scientific disciplines to enjoy the sort of success that scientific disciplines do (i.e., explanatory, predictive, and instrumental success), then we should want to introduce and make use of scientific methods in those non-scientific fields. For instance, Wesley Buckwalter and John Turri argue that the application of methods from the social sciences in philosophy (also known as "experimental philosophy") has been quite successful. As they put it, "Experimental, observational, and statistical techniques have significantly contributed to research in epistemology, action theory, ethics, philosophy of language, and philosophy of mind" (Buckwalter and Turri 2018, 282). Similarly, Moti Mizrahi argues that the introduction of methods from data science into the philosophy of logic particularly (Mizrahi 2019) and philosophy generally (Mizrahi 2018b, 48) might bring to academic philosophy the sort of success enjoyed by the sciences. And Ibo Van De Poel (2020, 231–244) argues that, if they want to make academic philosophy "societally relevant," academic philosophers need to incorporate the synthetic methods of designers as well as techniques of experimentation into philosophical inquiry.

6. OVERVIEW OF THIS VOLUME

This edited volume arises from an exchange between several scholars over the pages of the *Social Epistemology Review and Reply Collective* generated by the publication of my "The Scientism Debate: A Battle for the Soul of Philosophy?" (see chapter 2 of this volume). I am very grateful to those who have contributed to that exchange and to this collection. I am also grateful to the Editor, James H. Collier, for inviting me to edit this collection.

The contributions to this collection share in common the view that the term "scientism" should not be weaponized. Instead, the term should remain neutral and any view labeled "scientism" (or some variation thereof) should be evaluated on its own merits. In other words, there is a substantive and important debate to be had about whether scientific knowledge, methods, practices, and ways of knowing are superior to non-scientific ones, and if so, in what respects. Furthermore, if academic philosophy is an essentially a priori form of inquiry, then does scientism pose an existential threat to philosophy? Or is scientism a way to make philosophy as successful as science?

Recently, there has been increased interest in scientism among philosophers, as evidenced by the publication of edited books, such as Boudry and Pigliucci (2017) and de Ridder et al. (2018). Both books, however, tend to take a strong version of scientism as their target of criticism. For example, in de Ridder et al. (2018), scientism is defined at the outset as "the view that *only* science can provide us with knowledge or rational belief, that *only* science can tell us what exists, and that *only* science can effectively address our moral and existential questions" (van Woudenberg et al. 2018, 1; emphasis added).[9] Similarly, in Boudry and Pigliucci (2017, 2), scientism is defined at the outset as "the conviction that the methods of science are the *only* worthwhile modes of inquiry, and will eventually supplant all others" (emphasis added). In so doing, both books erect, and then take down, a strawman. They fail to consider more sophisticated and plausible versions of scientism, such as *Weak Scientism* (namely, the view that, of all the knowledge we have, scientific knowledge is the *best*—though not the only—kind of knowledge). By contrast, this edited collection contains chapters that critically examine more plausible versions of scientism, thereby allowing for the question of scientism and its alleged threat to philosophy to be genuinely open to debate.

In chapter 2, I consider the possibility that, beneath the surface of the scientism debate in academic philosophy, there might be real concerns that academic philosophers have about the future of their discipline, both as a field of study and as a field of inquiry. Taking an empirical approach, I set out to test two hypotheses about the scientism debate in philosophy. According to the first hypothesis, the scientism debate is fundamentally about philosophy as a field of study (i.e., a subject for undergraduate students to major in).

That is, academic philosophers find scientism threatening because they see it as a threat to the future of philosophy as a major in colleges and universities. According to the second hypothesis, the scientism debate is fundamentally about philosophy as a field of inquiry (i.e., a subject for scholars to work in). That is, academic philosophers find scientism threatening because they see it as a threat to the soul or essence of philosophy as an a priori discipline. I find some empirical evidence suggestive of a link between the introduction of empirical methods into philosophy and concerns about scientism among philosophers that is worthy of further investigation.

In chapter 3, Amanda Bryant meticulously surveys the sort of assumptions that would have to be made for scientism to pose a threat to academic philosophers. Bryant argues that weak and even strong versions of scientism are not inherently threatening to academic philosophers unless a few substantive epistemological, methodological, and value-theoretic assumptions are made.

In chapter 4, Ian James Kidd invites us to think carefully about the metaphilosophical conflicts that lie just beneath the surface of philosophical debates about scientism. Drawing on the work of Mary Midgley, Kidd argues that we need to esteem science in the right way. This requires having a conception of philosophy in contradistinction to science that does justice to both.

In chapter 5, Petri Turunen, Ilkka Pättiniemi, Ilmari Hirvonen, Johan Hietanen, and Henrik Saarinen distinguish between narrow and broad versions of Weak Scientism. They defend a broad version of Weak Scientism by appealing to what they call "epistemic opportunism." They argue that epistemic opportunism explains why scientists are willing to adopt any method that works, which is how science ought to be practiced.

In chapter 6, Luana Poliseli and Federica Russo consider the consequences of scientism for philosophical methodology. In particular, they discuss the field of Philosophy of Science in Practice (PSP) as a successful example of the use of diverse methodologies, including empirical methods, to address philosophical questions about science. From the successful introduction of empirical and empirically informed methodologies into PSP, Poliseli and Russo argue against hierarchies of knowledge and for methodological diversity in philosophy.

In chapter 7, Catherine Wilson focuses on scientism as an expression of moral critiques of the practice of science. The first critique is what Wilson calls "misappropriation," which is the claim that too much money is spent on science at the expense of other worthwhile projects. The second critique is what Wilson calls "no boundaries," which is the claim that the practical goals of science, namely, controlling and manipulating nature, are not necessarily conducive to the well-being of humankind.

In chapter 8, Ann-Sophie Barwich approaches the scientism debate as it plays out on the borderlines between philosophy and cognitive science. In particular, through an examination of the view known as "panpsychism" in philosophy of mind, Barwich argues that an unwanted consequence of philosophical arguments for panpsychism is a general skepticism about science. For Barwich, avoiding this unwanted consequence requires a new image of science that can facilitate collaboration between the sciences and the humanities.

In chapter 9, I conclude this collection with some reflections on the future of academic philosophy. If academic philosophy were to change from an essentially a priori discipline to a discipline that looks more like science, would that "spell shipwreck for philosophy itself" (Haack 2017, 43), as some fear, or would it make philosophy as successful and progressive as science, as some proponents of scientism argue?

NOTES

1. For a detailed conceptual map of scientism, see Peels (2018).

2. For a recent discussion of scientism in the domains of medical science and clinical practice, see Loughlin et al. (2013).

3. Cf. Paul Hoyningen-Huene (2013) on science as systematic knowledge.

4. See Shiping Hua (1995, 15–22) on "empirical scientism" versus "materialistic scientism."

5. See D. M. Armstrong (1978, 261) on materialism or physicalism as a subspecies of naturalism.

6. See Fink (2006) for further discussion. Cf. Burch (2016).

7. Cf. Wilfrid Sellars' (1997, 83) claim, paraphrasing Protagoras, that "science is the measure of all things; of what is that it is and of what is not that it is not."

8. Hietanen et al. (2020, 526) add another dimension to epistemological scientism, namely, narrow versus broad epistemological scientism. Narrow epistemological scientism is a thesis about the natural sciences in particular, whereas broad epistemological scientism is a thesis about the sciences in general.

9. Although, in the same volume, Peels (2018) goes on to identify several varieties of scientism, including strong and weak versions.

REFERENCES

Andrews, Stephen P. 1871. *The Primary Synopsis of Universology and Alwato: The New Scientific Universal Language.* New York: Dion Thomas, 141 Fulton St.

Armstrong, D. M. 1978. "Naturalism, Materialism, and First Philosophy." *Philosophia* 8 (2–3): 261–276.

Atkins, Peter W. 1995. "Science as Truth." *History of the Human Sciences* 8 (2): 97–102.

Barbour, Ian G. 2001. "Science and Scientism in Huston Smith's *Why Religion Matters.*" *Zygon* 36 (2): 207–214.

Beale, Jonathan. 2017. "Wittgenstein's Anti-scientistic Worldview." In *Wittgenstein and Scientism*, edited by J. Beale and I. J. Kidd, 59–80. New York: Routledge.

Bell, David R. 1962. "Marx, Sartre and Marxism." *Manchester Literary and Philosophical Society: Memoirs and Proceedings* 104 (1961–62): 47–64.

Beltramini, Enrico. 2019. "Philosophy of Management Between Scientism and Technology." *Philosophy & Technology* 32 (3): 535–548.

Bernstein, Howard R. 1980. "Conatus, Hobbes, and the Young Leibniz." *Studies in History and Philosophy of Science Part A* 11 (1): 25–37.

Boudry, Maarten and Massimo Pigliucci. 2017. "Introduction." In *Science Unlimited? The Challenges of Scientism*, edited by M. Boudry and M. Pigliucci, 1–9. Chicago: The University of Chicago Press.

Bourget, David and David Chalmers. 2014. "What Do Philosophers Believe?" *Philosophical Studies* 170 (3): 465–500.

Braithwaite, R. B. 1953. *Scientific Explanation: A Study of the Function of Theory, Probability and Law in Science.* New York: Cambridge University Press.

Buckwalter, Wesley and Turri, John. 2018. "Moderate Scientism in Philosophy." In *Scientism: Prospects and Problems*, edited by J. de Ridder, R. Peels, and R. van Woudenberg, 280–300. New York: Oxford University Press.

Burch, Matthew. 2016. "Religion and Scientism: A Shared Cognitive Conundrum." *International Journal for Philosophy of Religion* 80 (3): 225–241.

Churchland, Patricia, S. 2011. *Braintrust: What Neuroscience Tells Us about Morality.* Princeton, NJ: Princeton University Press.

Davenport, Russell W. 1955. *The Dignity of Man.* New York: Harper and Brothers.

Day, Henry. N. 1870. "President McCosh's Logic." *The New Englander* 29 (3): 511–530.

Day, Henry. N. 1872. *The Science of Aesthetics or The Nature, Kinds, Laws, and Uses of Beauty.* New York: G. P. Putnam's Sons, 182 Fifth Ave.

Dupré, John. 1988. "Materialism, Physicalism, and Scientism." *Philosophical Topics* 16 (1): 31–56.

Dupré, John. 2001. *Human Nature and the Limits of Science.* Oxford: Oxford University Press.

Fink, Hans. 2006. "Three Sorts of Naturalism." *European Journal of Philosophy* 14 (2): 202–221.

Gasparatou, Renia. 2017. "Scientism and Scientific Thinking: A Note on Science Education." *Science & Education* 26 (7–9): 799–812.

Gulick, Addison. 1955. "Evolutionary Naturalism alias 'Scientism'." *The Scientific Monthly* 80 (6): 392–393.

Haack, Susan. 2007. *Defending Science--Within Reason: Between Scientism and Cynicism.* Amherst: Prometheus Books.

Haack, Susan. 2012. "Six Signs of Scientism." *Logos & Episteme* 3 (1): 75–95.

Haack, Susan. 2017. "The Real Question: Can Philosophy Be Saved?" *Free Inquiry* 37 (6): 40–43.

Harrison, Peter. 2018. "Naturalism and the Success of Science." *Religious Studies* 56 (2): 274–291.

Haught, John F. 2005. "Science and Scientism: The Importance of a Distinction." *Zygon* 40 (2): 363–368.

Hayek, F. A. 1942. "Scientism and the Study of Society." *Economica* 9 (35): 267–291.

Hempel, Carl G. 1973. "Science Unlimited?" *Annals of the Japan Association for Philosophy of Science* 4 (3): 187–202.

Hietanen, J., Turunen, P., Hirvonen, I., Karisto, J., Pättiniemi, I., and Saarinen, H. 2020. "How Not to Criticise Scientism." *Metaphilosophy* 51 (4): 522–547.

Hoyningen-Huene, Paul. 2013. *Systematicity: The Nature of Science*. New York: Oxford University Press.

Hua, Shiping. 1995. *Scientism and Humanism: Two Cultures in Post-Mao China (1978-1989)*. Albany: State University of New York Press.

Johnson, Phillip E. 1991/2010. *Darwin on Trial*. Third Edition. Downers Grove, IL: InterVarsity Press.

Kidd, Ian James. 2016. "How Should Feyerabend Have Defended Astrology?" *Social Epistemology Review and Reply Collective* 5 (6): 11–17.

Kitcher, Philip. 1991. "Seeing in the Dark." *Nature* 354 (6349): 118–119.

Kitcher, Philip. 2012. "The Trouble with Scientism: Why History and the Humanities are also a Form of Knowledge." *The New Republic*, May 4, 2012. https://newre-public.com/article/103086/scientism-humanities-knowledge-theory-everything-arts-science.

Kitcher, Philip. 2017. "The Trouble with Scientism: Why History and the Humanities are Also a Form of Knowledge." In *Science Unlimited? The Challenges of Scientism*, edited by M. Boudry and M. Pigliucci, 109–120. Chicago: The University of Chicago Press.

Ladyman, James. 2011. "The Scientistic Stance: The Empirical and Materialist Stances Reconciled." *Synthese* 178 (1): 87–98.

Ladyman, James. 2002. *Understanding Philosophy of Science*. New York: Routledge.

Ladyman, James, Don Ross, David Spurrett, and John Collier. 2007. *Every Thing Must Go: Metaphysics Naturalized*. New York: Oxford University Press.

Loughlin, Michael, George Lewith, and Torkel Falkenberg. 2013. "Science, Practice and Mythology: A Definition and Examination of the Implications of Scientism in Medicine." *Health Care Analysis* 21 (2): 130–145.

Maddy, Penelope. 2005. "Three Forms of Naturalism." In *The Oxford Handbook of Philosophy of Mathematics and Logic*, edited by S. Shapiro, 437–459. New York: Oxford University Press.

McDowell, John. 1998. *Mind, Value, and Reality*. Cambridge, MA: Harvard University Press.

Midgley, Mary. 1992. *Science as Salvation: A Modern Myth and Its Meaning*. London: Routledge.

Mizrahi, Moti. 2017a. "What's so Bad about Scientism?" *Social Epistemology* 31 (4): 351–367.

Mizrahi, Moti. 2017b. "In Defense of Weak Scientism: A Reply to Brown." *Social Epistemology Review and Reply Collective* 6 (2): 9–22.

Mizrahi, Moti. 2018a. "More in Defense of Weak Scientism: Another Reply to Brown." *Social Epistemology Review and Reply Collective* 7 (4): 7–25.

Mizrahi, Moti. 2018b. "Weak Scientism Defended Once More." *Social Epistemology Review and Reply Collective* 7 (6): 41–50.

Mizrahi, Moti. 2018c. "Why Scientific Knowledge is Still the Best." *Social Epistemology Review and Reply Collective* 7 (9): 18–32.

Mizrahi, Moti. 2019. "What Isn't Obvious about 'Obvious': A Data-Driven Approach to Philosophy of Logic." In *Advances in Experimental Philosophy of Logic and Mathematics*, edited by A. Aberdein and M. Inglis, 201–224. London: Bloomsbury.

Neurath, Otto. 1987. "Unified Science and Psychology." In *Unified Science*, edited by B. F. McGuinness, 1–23. Dordrecht: D. Reidel Publishing Co.

Ney, Alyssa. 2018. "Physicalism, Not Scientism." In *Scientism: Prospects and Problems*, edited by J. de Ridder, R. Peels, and R. van Woudenberg, 258–279. New York: Oxford University Press.

Nussbaum, Martha, C. 2010. *Not for Profit: Why Democracy Needs the Humanities*. Princeton, NJ: Princeton University Press.

Olson, Richard G. 2008. *Science and Scientism in Nineteenth-Century Europe*. Chicago: University of Illinois Press.

Peels, Rik. 2017. "Ten Reasons to Embrace Scientism." *Studies in History and Philosophy of Science Part A* 63 (June): 11–21.

Peels, Rik. 2018. "A Conceptual Map of Scientism." In *Scientism: Prospects and Problems*, edited by J. de Ridder, R. Peels, and R. van Woudenberg, 28–56. New York: Oxford University Press.

Pennock, Robert T. 1996. "Naturalism, Evidence and Creationism: The Case of Phillip Johnson." *Biology and Philosophy* 11 (4): 543–559.

Pike, James. A. and John McGill Krumm. 1954. *Roadblocks to Faith: The Believer Answers the Skeptic*. New York: Morehouse-Gorham.

Psillos, Stathis. 1999. *Scientific Realism: How Science Tracks Truth*. London: Routledge.

Rasmusson, Richard H. 1954. "The Preacher Talks to the Man of Science." *The Scientific Monthly* 79 (6): 392–394.

de Ridder, Jeroen. 2014. "Science and Scientism in Popular Science Writing." *Social Epistemology Review and Reply Collective* 3 (12): 23–39.

Rosenberg, Alexander. 2011. *The Atheist's Guide to Reality: Enjoying Life Without Illusions*. New York: W. W. Norton.

Rosenberg, Alexander. 2017. "Strong Scientism and Its Research Agenda." In *Science Unlimited? The Challenges of Scientism*, edited by M. Boudry and M. Pigliucci, 203–223. Chicago: The University of Chicago Press.

Russell, Bertrand. 1946. *History of Western Philosophy, and Its Connection with Political and Social Circumstances from the Earliest Times to the Present Day*. London: Allen and Unwin.

Salmon, Merrilee H. 2013. *Introduction to Logic and Critical Thinking*. Sixth Edition. Boston, MA: Wadsworth.

Sellars, Wilfrid. 1997. *Empiricism and the Philosophy of Mind. With an Introduction by Richard Rorty and a Study Guide by Robert Brandom.* Cambridge, MA: Harvard University Press.

Sharlin, Harold I. 1976. "Herbert Spencer and Scientism." *Annals of Science* 33 (5): 457–465.

Sorell, Tom. 1991. *Scientism: Philosophy and the Infatuation with Science.* London: Routledge.

Special to *The New York Times*. 1907. "Soul has Weight, Physician Thinks; Dr. Macdougall of Haverhill Tells of Experiments at Death. Loss to Body Recorded; Scales Showed an Ounce Gone in One Case, He Says--Four Other Doctors Present." *The New York Times Archives.* Accessed on October 31, 2018. https://www.nytimes.com/1907/03/11/archives/soul-has-weight-physician-thinks-dr-macdougall-of-haverhill-tells.html.

Spencer, Herbert. 1873. *Social Statics; or, the Conditions Essential to Human Happiness Specified, and the First of Them Developed.* New York: D. Appleton and Co.

Stenmark, Mikael. 1997. "What is Scientism?" *Religious Studies* 33 (1): 15–32.

Stenmark, Mikael. 2004. *How to Relate Science and Religion: A Multidimensional Model.* Grand Rapids, MI: Wm. B. Eerdmans Publishing Co.

Stenmark, Mikael. 2016. *Scientism: Science, Ethics, and Religion.* New York: Routledge.

Van De Poel, Ibo. 2020. "Should Philosophers Begin to Employ New Methods if They Want to Become More Societally Relevant?" In *Philosophy in the Age of Science? Inquiries into Philosophical Progress, Method, and Societal Relevance*, edited by Julia Hermann, Jeroen Hopster, Wouter Kalf, and Michael Klenk, 231–244. London: Rowman & Littlefield.

Van Woudenberg, R., Peels, R., and de Ridder, J. 2018. "Introduction: Putting Scientism on the Philosophical Agenda." In *Scientism: Prospects and Problems*, edited by J. de Ridder, R. Peels, and R. van Woudenberg, 1–36. New York: Oxford University Press.

Werkmeister, W. H. 1959. "Scientism and the Problem of Man." *Philosophy East and West* 9 (1/2): 20–21.

Whitney, William Dwight (Ed.). 1890. *The Century Dictionary: An Encyclopedic Lexicon of the English Language.* Vol. V. Edited by William Dwight Whiteny. New York: The Century Co.

Willey, Angela. 2017. "Engendering New Materializations: Feminism, Nature, and the Challenge of Disciplinary Proper Objects." In *The New Politics of Materialism: History, Philosophy, Science*, edited by S. Ellenzweig and J. Zammito, 131–153. New York: Routledge.

Williams, R. N. 2015. "Introduction." In *Scientism: The New Orthodoxy*, edited by R. N. Williams and D. N. Robinson, 1–22. New York: Bloomsbury.

Wilson, Catherine. 2017. "Materialism, Old and New, and the Party of Humanity." In *The New Politics of Materialism: History, Philosophy, Science*, edited by S. Ellenzweig and J. Zammito, 111–130. New York: Routledge.

Chapter 2

The Scientism Debate
A Battle for the Soul of Philosophy?
Moti Mizrahi

What is the scientism debate in philosophy *really* about? Is the "scientism" charge made by academic philosophers against those who value science more than philosophy the battle cry of those philosophers who seek to defend their territory? If so, what are academic philosophers concerned about? Does scientism pose a threat to academic philosophy? In this chapter, I consider these questions by taking an empirical approach, borrowing statistical, computational, and corpus-based techniques from data science and computational linguistics. I set out to test two hypotheses about the scientism debate in philosophy empirically. According to the first hypothesis, the scientism debate is fundamentally about academic philosophy as a field of study (i.e., a subject for undergraduate students to major in). According to this hypothesis, academic philosophers find scientism threatening because they see it as a threat to the future of philosophy as a major in colleges and universities. According to the second hypothesis, the scientism debate is fundamentally about academic philosophy as a field of inquiry (i.e., an area for scholars to work in). According to this hypothesis, academic philosophers find scientism threatening because they see it as a threat to the soul or essence of philosophy as an a priori field of inquiry. I find some empirical evidence suggestive of a link between the introduction of empirical methods into academic philosophy and concerns about scientism among academic philosophers that is worthy of further investigation, or so I think.

1. THE PHILOSOPHER WHO CRIED "SCIENTISM"

The scientism debate in academic philosophy may seem rather fierce to the casual observer.[1] In the face of anyone (mostly prominent scientists, especially physicists, such as Steven Weinberg, Stephen Hawking, Lawrence Krauss, and Neil deGrasse Tyson) who is dismissive of philosophy, some academic philosophers cry "scientism."[2] There are good reasons why "scientism" is such a hot-button issue for academic philosophers, however, for the scientism debate gets at the heart of what philosophers do professionally, namely, their teaching and their research. As far as teaching is concerned, scientism may be perceived as a threat to the sort of teaching that academic philosophers typically do because it is commonly (but mistakenly) thought that degrees in science are in demand and can lead to successful careers, whereas degrees in philosophy are not and can only lead to unemployment (see, e.g., Shapiro 2017). As Graham Oddie (2006, 255) recounts, Alan Musgrave once told him that he "was doing something a bit foolish—passing up the chance for a degree (and a lucrative career) in law, for a degree (and almost certain unemployment) in philosophy." In other words, academic philosophers might think that scientism poses a threat to them as teachers because it somehow implies that philosophy has no valuable skills to impart to students. For this reason, some academic philosophers find it necessary to argue that philosophy can teach students something of value. For example, Martha Nussbaum argues that the arts and humanities can teach students skills that STEM fields cannot teach them, such as critical thinking, reasoning, and problem-solving skills (Nussbaum 2010, xvii–xviii).

As far as research is concerned, scientism may be perceived as a threat to the sort of research that academic philosophers typically do because it advocates for the use of empirical methods of observation, experimentation, and the like, whereas philosophers are typically content with armchair reflection. As Patricia Churchland (2011, 4) puts it, "philosophy and science are working the same ground, and [empirical] evidence should trump armchair reflection." In other words, academic philosophers might think that scientism poses a threat to them as researchers because it somehow implies that philosophy has no valuable contributions to make to the advancement of knowledge unless it adopts the empirical methods of the sciences. For this reason, some academic philosophers find it necessary to defend the traditional methods of philosophy against any attempt to introduce empirical methods into academic philosophy. For example, Jennifer Nagel defends the method of making intuitive judgments about hypothetical cases of philosophical interest (the so-called "method of cases") from experimental results suggesting that intuitive judgments in response to hypothetical cases are not as reliable as many philosophers tend to think (Nagel 2012).[3]

All of this suggests to me two things that the scientism debate in academic philosophy might *really* be about. First, the scientism debate is fundamentally a battle for the future of academic philosophy as a discipline or *field of study* (i.e., a subject for college students to major in): as teachers, academic philosophers seek to defend their territory from invading scientists by communicating to students the value of philosophy for their education and professional success. This is hypothesis 1: academic philosophers find scientism threatening because they see it as a threat to the future of philosophy as a major in colleges and universities. For instance, Massimo Pigliucci (2018) thinks that scientism is "a threat to every other discipline, including philosophy." Second, the scientism debate is fundamentally a battle for the soul or essence of academic philosophy as an area or *field of inquiry* (i.e., an area for scholars to work in): as scholars, academic philosophers seek to defend their territory from invading scientists by justifying the traditional methods of philosophical inquiry and resisting attempts to introduce the empirical methods of the sciences into academic philosophy. This is hypothesis 2: academic philosophers find scientism threatening because they see it as a threat to the soul or essence of philosophy as an a priori field of inquiry. For instance, Susan Haack (2017, 43) thinks that "the rising tide of scientistic philosophy [. . .] spells shipwreck for philosophy itself."[4]

Rather than reflect on these two hypotheses while sitting in an armchair, I set out to test them empirically, by borrowing corpus-based techniques from data science and computational linguistics, and thereby demonstrating the usefulness of empirical methods to philosophical (and/or metaphilosophical) inquiry. In this chapter, I describe how I tested the aforementioned hypotheses empirically and I report the results of my empirical study. Finally, I discuss the implications of the results of this empirical study for the scientism debate in academic philosophy.

2. METHODS AND RESULTS

For the purpose of this empirical study, then, my research question is this: What is the scientism debate in academic philosophy fundamentally about? As mentioned above, there are two hypotheses that will be tested empirically in this study. First, the scientism debate is fundamentally about philosophy as a field of study (i.e., a subject for undergraduate students to major in):

H1: Academic philosophers find scientism threatening because they see it as a threat to the future of philosophy as a major in colleges and universities.

Second, the scientism debate is fundamentally about philosophy as a field of inquiry (i.e., an area for scholars to work in):

H2: Academic philosophers find scientism threatening because they see it as a threat to the soul or essence of philosophy as an a priori field of inquiry.

It is important to note that these hypotheses are not to be read as universal generalizations (cf. de Ridder 2019). That is, H1 is not to be construed as the claim that all academic philosophers, without exception, find scientism threatening to the future of academic philosophy as a field of study. Similarly, H2 is not to be construed as the claim that all academic philosophers, without exception, find scientism threatening to the essence of academic philosophy as an a priori field of inquiry. Instead, both hypotheses should be read as statistical generalizations that can be subjected to empirical and statistical testing. Accordingly, on H1, academic philosophers find scientism threatening to academic philosophy as a field of study more often than not. In other words, if we were to pick an academic philosopher at random, that academic philosopher would be more likely than not to say that scientism poses a threat to academic philosophy as a field of study. Similarly, on H2, academic philosophers find scientism threatening to academic philosophy as an essentially a priori field of inquiry more often than not. In other words, if we were to pick an academic philosopher at random, that academic philosopher would be more likely than not to say that scientism poses a threat to academic philosophy as an essentially a priori field of inquiry.

Indeed, using computational and corpus-based techniques from data science and computational linguistics, we can find some empirical support for the claim that academic philosophers are more likely than not to harbor negative sentiments against scientism. In particular, we can run a sentiment analysis, also known as "opinion mining" (Liu 2017), on a random sample of philosophical texts about scientism to find out whether academic philosophers generally express positive or negative sentiments toward—or have positive or negative opinions about—scientism in their published works. To do this, I searched through PhilPapers (https://philpapers.org) for published articles with the word "scientism" in the title of the article. I selected the first twenty-five journal articles that came up in the search results on PhilPapers. These journal articles are the following:

1. "Materialism, Physicalism, and Scientism" by John Dupré
2. "Pragmatic Philosophy of Science and the Charge of Scientism" by Peter T. Manicas
3. "Naturalism, Scientism and the Independence of Epistemology" by James Maffie
4. "Scientism and Society" by J. W. N. Watkins
5. "What is Scientism?" by Mikael Stenmark
6. "Scientism and the Problem of Man" by W. H. Werkmeister
7. "Challengers of Scientism Past and Present: William James and Marilynne Robinson" by James Woelfel

8. "On the Poverty of Scientism, Or: The Ineluctable Roughness of Rationality" by Murray Code
9. "Scientism in the Arts and Humanities" Roger Scruton
10. "The Folly of Scientism" by Austin L. Hughes
11. "Nanotechnology and the Developing Critique of Scientism" by Lee-Anne Broadhead and Sean Howard
12. "Scientism, Social Praxis, and Overcoming Metaphysics: A Debate Between Logical Empiricism and the Frankfurt School" by Andreas Vrahimis
13. "Scientism and Pseudoscience: A Philosophical Commentary" by Massimo Pigliucci
14. "Scientism as a Social Response to the Problem of Suicide" by Scott J. Fitzpatrick
15. "Aesthetics, Scientism, and Ordinary Language: A Comparison between Wittgenstein and Heidegger" by Andreas Vrahimis
16. "Scientism and Scientific Thinking" by Renia Gasparatou
17. "Scientism, Deconstruction, and Nihilism" by Nicholas Capaldi
18. "Six Signs of Scientism" by Susan Haack
19. "Ten Reasons to Embrace Scientism" by Rik Peels
20. "Scientism, Pragmatism, and the Fate of Philosophy" by Kai Nielsen
21. "Scientism and Technology as Religions" by Rustum Roy
22. "Doing Away with Scientism" by Ian James Kidd
23. "Scrutinizing Scientism from a Hermeneutic Point of View" by Dimitri Ginev
24. "Against Scientism, for Personhood" by Christian Perring
25. "Wittgenstein, Mind and Scientism" by Warren Goldfarb

In order to find out whether these articles contain mostly positive or negative sentiments toward (or opinions about) scientism, I ran a sentiment analysis on the titles and abstracts of these twenty-five articles using the Azure Machine Learning add-in in Microsoft Excel (Microsoft Office 365). Azure Machine Learning is a free analytics tool that uses Natural Language Processing (NLP) to run analyses, such as sentiment analysis, on unstructured text. A sentiment analysis "is a process of automatically extracting opinions or emotions from text, especially in user-generated textual content. Sentiment analysis is considered a classification task which classifies text into positive, negative, or neutral classes" (Kumar and Harish 2020, 1122). The Azure Machine Learning text sentiment analysis uses the Multi-Perspective Question Answering (MPQA) Subjectivity Lexicon (http://mpqa.cs.pitt.edu/lexicons/subj_lexicon/), which is a commonly used subjectivity lexicon in NLP. The MPQA Subjectivity Lexicon includes 5,097 negative words and 2,533 positive words with strong and weak polarity annotations. As Theresa Wilson, Janyce Wiebe, and Paul Hoffman (2005, 348) explain:

> The *positive* tag is for positive emotions (*I'm happy*), evaluations (*Great idea!*), and stances (*She supports the bill*). The *negative* tag is for negative emotions (*I'm sad*), evaluations (*Bad idea!*), and stances (*She's against the bill*). [. . .] The *neutral* tag is used for all other subjective expressions: those that express a different type of subjectivity such as speculation, and those that do not have positive or negative polarity (emphasis in original).[5]

Accordingly, the output of a sentiment analysis performed by the Azure Machine Learning algorithm includes the sentiment tags of "positive," "negative," and/or "neutral," as well as their associated scores between zero and one. A score close to zero gets a "negative" tag, a score close to one gets a "positive" tag, and a score approximately midpoint between zero and one gets a "neutral" tag.

Accordingly, if we run a sentiment analysis on the text from the titles and abstracts of the aforementioned twenty-five published articles with the word "scientism" in the title, we can find out whether the authors of these published articles express mostly positive, negative, or neutral opinions toward scientism. Since our sample of published articles with the word "scientism" in the title was generated randomly (i.e., presumably, any published article with the word "scientism" in the title has had a roughly equal chance of making it into the search results on PhilPapers), we can be quite confident that the results of the sentiment analysis will be fairly representative of the sentiments of academic philosophers toward scientism rather generally.

The results of the sentiment analysis are as follows. Of the twenty-five articles on the list above, the Azure Machine Learning algorithm tagged seventeen as "negative" (68%), five as "positive" (20%), and three as "neutral" (12%). The mean score of the negative articles is 0.07, the mean score of the positive articles is 0.87, and the mean score of the neutral articles is 0.47. These results suggest that, for the most part, articles on scientism written by academic philosophers tend to contain mostly negative, rather than positive (or neutral), sentiments about scientism. In other words, if we were to pick at random a journal article about scientism written by an academic philosopher, that article is more likely to contain negative, rather than positive (or neutral), opinions about scientism.

Now, let's return to the two hypotheses we would like to test empirically! If H1 were true, we would expect academic philosophers to feel threatened when they lose students to STEM majors. In other words, if more students choose to major in STEM fields instead of philosophy, then academic philosophers might feel more threatened by scientism. Conversely, if more students choose to major in philosophy instead of STEM fields, then academic philosophers might feel less threatened by scientism. In statistical terms, if H1 were true, we would expect to find a positive linear relationship between the proportion of students that choose STEM over philosophy and how concerned academic philosophers are about scientism.

So now the question is how to find out whether there is a relationship between undergraduate students choosing to major in STEM over philosophy and academic philosophers being concerned about scientism. Data on the subjects undergraduate students choose to major in is relatively easy to come by. The Institute of Education Sciences' (IES) National Center for Education Statistics (NCES) provides data on bachelor's degrees conferred by postsecondary institutions in the United States through its Integrated Postsecondary Education Data System (IPEDS). Since my focus is the scientism debate in academic philosophy, I have looked at data on Philosophy and Religious Studies bachelor's degrees in comparison with bachelor's degrees in STEM, specifically, Biological and Biomedical Sciences; Engineering, Mathematics, and Statistics; Physical Sciences; and Psychology (these are NCES' groupings, not mine), which I have taken from the NCES' 2019 Digest of Education Statistics (available here: https://nces.ed.gov/programs/digest/d20 /tables/dt20_322.10.asp). The data relevant to this empirical study are summarized in table 2.1.

A correlation analysis indicates that there is a negative correlation between the percentage of STEM bachelor's degrees and the percentage of Philosophy and Religious Studies bachelor's degrees conferred by postsecondary institutions in the United States from 2005 to 2018 ($r = -0.96$). The Pearson correlation coefficient r can tell us about the linear relationship between two variables (positive or negative) and the strength of that relationship (the closer r is to 0, the weaker the linear relationship; the closer r is to -1, the stronger the negative relationship; the closer r is to 1, the stronger the positive relationship). Accordingly, the negative correlation between the percentage of STEM bachelor's degrees and the percentage of Philosophy and Religious Studies bachelor's degrees conferred by postsecondary institutions in the United States from 2005 to 2018 ($r = -0.96$) is a strong negative correlation. In addition, the results of a multiple regression analysis indicate that the percentage of STEM bachelor's degrees by year explained 93 percent of the variance in the percentage of Philosophy and Religious Studies bachelor's degrees conferred by postsecondary institutions in the United States from 2005 to 2018, $R^2 = 0.93$, $F(2, 11) = 84.085$, $p < .001$.[6] See figure 2.1.

It is a bit more challenging to gather data on how concerned academic philosophers are about scientism. Nonetheless, when scholars are concerned about something, they tend to write about it, both in academic journals (see, e.g., Bilgrami 2014) and books (see, e.g., Scruton 2014) as well as in

Table 2.1   Number and Percentage of Bachelor's Degrees in Philosophy and Religious Studies Compared to Bachelor's Degrees in STEM Conferred by Postsecondary Institutions in the United States from 2005 to 2018

Year	Total	STEM	STEM/Total	Philosophy and Religious Studies	Philosophy and Religious Studies/Total
2005	1,485,104	260,874	0.17	11,985	0.008
2006	1,601,399	269,990	0.16	11,969	0.007
2007	1,563,069	278,218	0.17	12,257	0.007
2008	1,601,399	284,210	0.17	12,448	0.007
2009	1,649,919	295,673	0.17	12,503	0.007
2010	1,716,053	309,133	0.18	12,830	0.007
2011	1,792,163	331,825	0.18	12,645	0.007
2012	1,840,381	349,332	0.18	12,792	0.006
2013	1,870,150	364,432	0.19	11,999	0.006
2014	1,894,969	377,225	0.19	11,071	0.005
2015	1,920,750	391,293	0.21	10,157	0.005
2016	1,956,114	404,645	0.21	9,711	0.004
2017	1,980,665	413,845	0.21	9,603	0.004
2018	2,012,854	421,708	0.21	9,615	0.004

Source: NCES.

non-academic venues (see, e.g., Schmidt 2018). Accordingly, if we look at how many publications in academic philosophy include a discussion of scientism, we will have a pretty good idea of how concerned academic philosophers are about scientism. In other words, the more concerned academic philosophers are about scientism, the more articles and book chapters they would write about scientism and its perceived threat to academic philosophy.

Such data can be mined from the JSTOR database using JSTOR's platform for mining textual data and creating datasets, namely, Constellate (https:// constellate.org/). I have used it to create a dataset of philosophy publications from the JSTOR database and searched for publications in which the term "scientism" occurs. Table 2.2 lists the number of publications in the Philosophy subject category on JSTOR that contain the term "scientism."[7]

The years in table 2.2 were selected to match the years in the data from NCES (see table 2.1).

As far as the percentage of STEM bachelor's degrees conferred by postsecondary institutions in the United States from 2005 to 2018 and philosophy publications that mention "scientism" are concerned, a correlation analysis indicates that there is a negative correlation between the percentage of STEM bachelor's degrees conferred by postsecondary institutions in the United States from 2005 to 2018 and the percentage of philosophy publications that contain the term "scientism" in the Philosophy subject category on JSTOR from 2005 to 2018 ($r = -0.33$). In addition, a multiple regression analysis did not find a significant regression

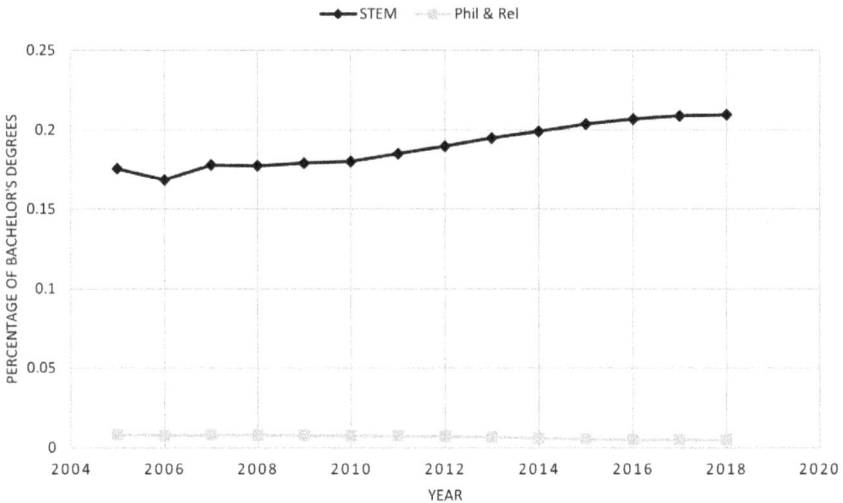

Figure 2.1 Percentage of Bachelor's Degrees in Philosophy and Religious Studies Compared to Bachelor's Degrees in STEM Conferred by Postsecondary Institutions in the United States from 2005 to 2018.

equation, $R^2 = 0.12$, $F(2, 11) = 0.813$, $p = .46$. This suggests that the percentage of STEM bachelor's degrees conferred by postsecondary institutions in the United States by year does not account for the variation in the number of publications that contain the term "scientism" in the Philosophy subject category on JSTOR.

As mentioned above, if H1 were true, we would expect to find a positive linear relationship between the proportion of students that choose STEM over philosophy and how concerned academic philosophers are about scientism. The results of correlation and regression analyses suggest that more students choose to major in STEM over Philosophy and Religious Studies. However, these results do not provide any indication that this trend in bachelor's degrees conferred makes academic philosophers more concerned about scientism such that they write more about it in their academic publications, since we did not find a positive linear relationship between the proportion of students that choose STEM over philosophy and the percentage of philosophy publications that mention scientism.

Now, if H2 were true, we would expect academic philosophers to feel more threatened by scientism when they think that the traditional methods of philosophical inquiry (such as the so-called "method of cases" and appealing to intuitions) begin to lose ground to empirical methods of investigation. In other words, if more scientific methods are being introduced into academic philosophy, then academic philosophers would feel more threatened by scientism, which would be reflected in an increased number of publications

Table 2.2   Number and Percentage of Publications that Contain the Term "Scientism" in the Philosophy Subject Category on JSTOR from 2005 to 2018

Year	Total	Scientism	Scientism/Total
2005	19,898	158	0.007
2006	19,327	166	0.008
2007	19,872	167	0.008
2008	19,821	169	0.008
2009	19,789	162	0.008
2010	19,748	145	0.007
2011	20,569	155	0.007
2012	20,817	179	0.008
2013	20,747	149	0.007
2014	19,702	151	0.007
2015	19,943	152	0.007
2016	11,717	101	0.008
2017	11,531	88	0.007
2018	3,357	26	0.007

Source: JSTOR.

that discuss scientism. In statistical terms, if H2 were true, we would expect to find a positive linear relationship between the number of philosophy publications that make use of empirical methods and how concerned academic philosophers are about scientism, as indicated by the number of philosophy publications that discuss scientism.

So now the question is how to find out whether there is a relationship between the popularity of empirical methods in academic philosophy and academic philosophers being concerned about scientism (as indicated by the number of articles and book chapters on scientism they publish). The conventional wisdom in academic philosophy is that the early years of the twenty-first century marked the advent of experimental philosophy (Knobe and Nichols 2017). Accordingly, if we find that there is a relationship between the increased popularity of experimental philosophy and concern among academic philosophers over scientism, then that would count as some positive evidence for H2.

Again, data on scientism and experimental philosophy in philosophy publications can be mined from the JSTOR database using JSTOR's platform for mining textual data and creating datasets, namely, Constellate (https://constellate.org/). I have used it to create a dataset of philosophy publications from the JSTOR database and searched for the term "scientism" and the phrase "experimental philosophy." Table 2.3 lists the number of publications that contain the term "scientism" as well as those that contain the phrase "experimental philosophy" mined from the Philosophy subject category in the JSTOR database. For the purpose of this analysis, I went as far back as 2008, since it is generally considered to be the year that marks the advent of experimental philosophy with the publication of Joshua Knobe and Shaun Nichols' "An Experimental Philosophy Manifesto" in Knobe and Nichols (2008).[8]

A correlation analysis indicates that there is a positive correlation between the number of philosophy publications in which the term "scientism" occurs and those in which the phrase "experimental philosophy" occurs ($r = 0.93$), which is a strong positive correlation. In addition, the results of a multiple regression analysis indicate that the number of "experimental philosophy" publications by year explained 95 percent of the variance in the percentage of "scientism" publications in philosophy from 2008 to 2018, $R^2 = 0.95$, $F(2, 8) = 83.842$, $p < .001$.[9] See figure 2.2.

The results of the multiple regression analysis indicated that the two predictors, namely, "experimental philosophy" publications and year, explained 95 percent of the variance, $R^2 = 0.95$, $F(2,8) = 83.842$, $p < .001$. It was found that "experimental philosophy" publications ($\beta_1 = 0.13$, $p < .001$) by year ($\beta_0 = -4.67$, $p = .006$) significantly predicted "scientism" publications.

As mentioned above, if H2 were true, we would expect to find a positive linear relationship between the number of philosophy publications that make use of empirical methods and how concerned academic philosophers are about scientism, as indicated by the number of philosophy publications that discuss scientism. The results of correlation and regression analyses suggest that the number of philosophy publications in which the term "scientism" occurs and the number of philosophy publications in which the phrase "experimental philosophy" occurs are positively correlated and that the variation in the former is explained by the variation in the latter over the years.

Table 2.3 Number of Publications that Contain the Term "Scientism" and the Phrase "Experimental Philosophy" in the Philosophy Subject Category on JSTOR from 2008 to 2018

Year	Total	Scientism	Experimental Philosophy
2008	19,821	169	935
2009	19,789	162	914
2010	19,748	145	934
2011	20,569	155	1048
2012	20,817	179	1009
2013	20,747	149	1057
2014	19,702	151	1018
2015	19,943	152	1070
2016	11,717	101	687
2017	11,531	88	700
2018	3,357	26	227

Source: JSTOR.

3. DISCUSSION

In the previous section, I reported the results of an empirical study, which was designed to test the following hypotheses:

H1: Academic philosophers find scientism threatening because they see it as a threat to the future of philosophy as a major in colleges and universities.

H2: Academic philosophers find scientism threatening because they see it as a threat to the soul or essence of philosophy as an a priori field of inquiry.

With respect to H1, the results of this empirical study do not lend any empirical support to H1. Contrary to what we would expect to find if H1 were true, we find a negative correlation between the percentage of STEM bachelor's degrees conferred by postsecondary institutions in the United States from 2005 to 2018 and the percentage of philosophy publications that contain the term "scientism" in the Philosophy subject category on JSTOR from 2005 to 2018.

With respect to H2, the data show a strong positive correlation between the number of philosophy publications that contain the term "scientism" and those that contain the phrase "experimental philosophy" ($r = 0.93$). This strong positive correlation between the number of philosophy publications in

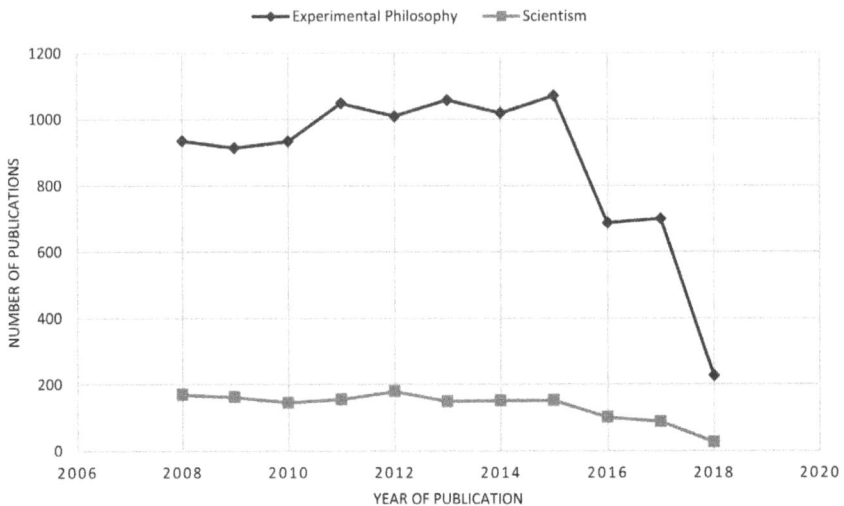

Figure 2.2   Number of Publications in which the Phrase "Experimental Philosophy" Occurs in Relation to those in which the Term "Scientism" Occurs in the Philosophy Subject Category on JSTOR from 2008 to 2018.

which the term "scientism" occurs and those in which the phrase "experimental philosophy" occurs is what we would expect to find if H2 were true. For, if H2 is true, we would expect academic philosophers to feel more threatened by scientism when they think that the traditional methods of philosophical investigation (such as the so-called "method of cases" and appealing to intuitions) begin to lose ground to empirical methods of investigation. This positive correlation and the result of a linear regression analysis, which indicates that the number of "experimental philosophy" publications predicts "scientism" publications in philosophy over the years, suggest a link between the introduction of empirical methods into academic philosophy and concerns about scientism among academic philosophers that is worthy of further investigation, or so I think.

Moreover, it looks like these results are in line with the results of other empirical studies on the use of empirical methods in academic philosophy. In one empirical study, Joshua Knobe (2015) compared two samples of papers on philosophy of mind: one sample of papers from 1960 to 1999 and another sample of papers from 2009 to 2013. Knobe (2015) found that 62 percent of the papers from the 1960 to 1999 sample used purely a priori methods, whereas only 12 percent of the papers from the 2009 to 2013 sample used purely a priori methods. This evidence leads Knobe (2015, 38) to conclude that there has been "a strong shift [in method] toward the use of systematic empirical data, including original experiments conducted by philosophers [that is, experimental philosophy]."

In another empirical study, Ashton and Mizrahi (2018b) test the view that philosophy is essentially an a priori discipline empirically. According to Ashton and Mizrahi (2018b, 62), "if philosophy is indeed *a priori*, and in the business of discovering necessary truths from the armchair, we would expect philosophers to advance mostly deductive, not inductive, arguments." Consistent with the view that philosophy is an a priori discipline, Ashton and Mizrahi (2018b) find that the proportion of philosophy papers in which deductive arguments are made is higher than that of philosophy papers in which inductive arguments are made. However, contrary to the view that philosophy is an a priori discipline, Ashton and Mizrahi (2018b) also find that the proportions of philosophy papers in which deductive arguments are made and those in which inductive arguments are made are converging over time and that the difference between the ratios of inductive and deductive arguments is declining over time. As Ashton and Mizrahi (2018b, 68–69) put it, their data suggest that "deductive arguments are gradually losing their status as the dominant form of argumentation in philosophy."

Both of these empirical studies, then, find trends that reveal a methodological shift in academic philosophy: a shift away from purely a priori or conceptual methods of investigation (so-called "armchair philosophy") toward a posteriori or empirical methods of investigation (so-called "experimental philosophy"). If this methodological shift in academic philosophy is really happening, and H2 is true, then we would expect to see an increase in the number of articles and book chapters in which academic philosophers discuss scientism.

Now, opponents of scientism, who think that science should not go beyond its proper boundaries, might be alarmed by this methodological shift from the armchair to the laboratory (Haug 2014, 1–26). They might characterize this shift as "scientific imperialism," which is "the tendency for a successful scientific idea to be applied far beyond its original home" (Dupré 2001, 16). For example, Tom Sorell (2017, 265) finds experimental philosophy "scientistic" insofar as "it objectionably treats natural science as the preferred body of results and methods for intellectual work of every kind."[10] While opponents of scientism want to resist the application of scientific ideas, methods, and practices in new domains, proponents of scientism think that it would actually be a good idea. For example, Wesley Buckwalter and John Turri (2018) argue for what they call "moderate scientism." Moderate scientism is the view that the empirical methods of the sciences can help test hypotheses and answer questions in non-scientific disciplines. In particular, Buckwalter and Turri argue for moderate scientism in academic philosophy, since the application of methods from the social sciences in philosophy (i.e., "experimental philosophy") has been quite successful. As they put it, "Experimental, observational, and statistical techniques have significantly contributed to research in epistemology, action theory, ethics, philosophy of language, and philosophy of mind" (Buckwalter and Turri 2018, 282). Similarly, I argue that the introduction of methods from data science into philosophy of logic particularly (Mizrahi 2019) and philosophy generally (Mizrahi 2018b, 48) might bring to academic philosophy the sort of success enjoyed by the sciences. And Ibo Van De Poel (2020, 231–244) argues that, if they want to make academic philosophy "societally relevant," academic philosophers need to incorporate the synthetic methods of designers as well as techniques of experimentation into philosophical inquiry.

Moreover, the charge of "scientific expansionism" (Stenmark 1997, 18) or "scientific imperialism" (Kitcher 2017, 110–112) leveled against anyone who advocates for the adoption of scientific methods in academic philosophy can be turned on its head quite easily. That is, opponents of scientism may see the introduction of scientific methods into academic philosophy as *expansionism*, that is, a misguided tendency to introduce scientific methods into territories they do not belong to traditionally, whereas proponents of scientism may see the resistance to scientific methods in academic philosophy as *territorialism*, that is, a misguided tendency to defend the so-called traditional territories of philosophy at all cost. For instance, against Tamara Browne's argument that an ethics review panel consisting of philosophers and bioethicists should oversee revisions to the *Diagnostic and Statistical Manual of Mental Disorders* (DSM) because such revisions require answering value-laden questions that "science cannot answer authoritatively" (Browne 2017, 188), Charlotte Blease (2017, 234) argues that, if we are to seek "the best interests of the patient," then "any such panel [. . .] would not be restricted to philosophers." Insisting that only philosophers can talk about values, Blease (2017) argues, amounts to philosophical territorialism.

As Blease (2017) also argues, the claim that science cannot answer value-laden questions rests on a rigid distinction between facts and values, the so-called fact/value dichotomy. That is, if science is all about facts, not values, then science cannot answer value-laden questions, or so the argument goes. There are at least two problems with this argument. First, the so-called fact/value dichotomy is a rather controversial distinction even among academic philosophers. For example, according to Hilary Putnam (2002, 19), "the fact/value dichotomy is, at bottom, not a *distinction* but a *thesis*, namely the thesis that 'ethics' is not about 'matters of fact'" (emphasis in original). When "we *disinflate* the fact/value dichotomy" in this way, Putnam (2002, 19) argues, what we get is an innocuous thesis about the different types of judgments one could make, such as ethical judgments, chemical judgments, sociological judgments, and so on. However, Putnam (2002, 19) argues, "*nothing metaphysical follows from the existence of a fact/value distinction in this (modest) sense*" (emphasis in original). The so-called Hume's Law or "*is-ought* gap," that is, that one cannot derive an *ought* from an *is*, which supposedly underpins the fact/value dichotomy, is also far from generally accepted among academic philosophers. For example, Philippa Foot (1958) argues that so-called thick ethical concepts, such as *rude,* show that one can derive an *ought* from an *is*. According to Foot, the word "rude" is normative, as opposed to descriptive, since it expresses condemnation in much the same way that the words "wrong" and "bad" do, and yet one can infer "this behavior is rude," which is a normative or evaluative conclusion, from "this behavior causes offense by indicating lack of respect," which is a descriptive premise (Foot 1958, 507–508).

Second, it is not the case that science is all about facts, not values. Rather, science is value-laden. For example, Thomas Kuhn famously argued that the following values guide scientists when they choose between competing theories: "accuracy, precision, scope, simplicity, fruitfulness, consistency, and so on" (Kuhn 2000, 251). That science is value-free, and all about facts, is an ideal that is not only far from actual scientific practice but also undesirable (Douglas 2009, 14). Accordingly, the argument that science cannot answer value-laden questions because it is all about facts, not values, rests on a problematic fact/value dichotomy and a false premise about science being allegedly free from values.[11]

Much like the charge of scientific expansionism, however, the charge of philosophical territorialism does not have a lot to stand on without reasons to believe that the territories one seeks to defend (or encroach upon) would benefit from resisting (or allowing) the introduction of new methods from outside those territories. For opponents of scientism, the reason for being territorial about philosophy seems to be the notion that philosophy is an essentially a priori discipline and should remain so; in other words, philosophy can only be practiced from the armchair, and not in the laboratory. But why should one think that philosophy is essentially an a priori form of inquiry? Why can it be done from the armchair only and not from the laboratory? Opponents of scientism who are territorial about philosophy need to provide principled answers to these questions. Simply saying that philosophy has always been that way, as some do (see, e.g., Sorell 2018), is not a satisfactory answer.[12] For proponents of scientism, on the other hand, the reason to embrace scientific methods in philosophy is their track record of success in the sciences. Since these methods have been applied successfully to test hypotheses and answer questions in the sciences, one can reasonably expect these methods to be successful, at least to some extent, when used to test hypotheses and answer questions in philosophy.[13]

In that respect, beyond testing hypotheses about the scientism debate in academic philosophy, this study serves as a demonstration of the usefulness of empirical methods to philosophical (and/or metaphilosophical) inquiry, as I have mentioned at the outset of this chapter. That is, the statistical, computational, and corpus-based methods used in this study can also be used to address other philosophical and metaphilosophical questions. For example, in several papers (Mizrahi 2021a, 2022), I use similar corpus-based methods to address the question of scientific progress in philosophy of science. For the most part, philosophers of science have dealt with the question of progress in science either by considering hypothetical scenarios and then using the intuitions elicited from those hypothetical scenarios as evidence for and/ or against philosophical accounts of scientific progress or by appealing to selected case studies from the history of science. Thus far, the application of these methods to the question of scientific progress has failed "to compel agreement" (Chalmers 2015, 25) among philosophers of science. Arguably, it is generally true that appealing to intuitions and appealing to case studies are methodologies that have failed "to compel agreement" (Chalmers 2015, 25) among academic philosophers, no matter the philosophical question to which these methods have been applied thus far, which is why some philosophers bemoan the lack of progress in philosophy (Chalmers 2015). It may also be why "Disagreement in philosophy is pervasive and irresoluble" (van Inwagen 2004, 332). Given the poor track record of these traditional methods in philosophy (namely, appealing to intuitions elicited from hypothetical cases and appealing to selected case studies from the history of science), then, it is reasonable to try new methods, such as corpus-based methods, to address philosophical questions.

4. CONCLUSION

As I have mentioned above, some academic philosophers who are opponents of scientism would consider a methodological shift from the armchair to the laboratory to be potentially disastrous to the future of philosophy as a field of inquiry. For instance, Susan Haack (2017, 43) thinks that "the rising tide of scientistic philosophy [. . .] spells shipwreck for philosophy itself." The results of this empirical study, which are consistent with the results of two other empirical studies, namely, Knobe (2005) and Ashton and Mizrahi (2018b), may point to such a methodological shift in philosophy, but they tell us nothing about whether such a shift would be good or bad for philosophy as a field of inquiry. Nevertheless, I think it is important to remember that attempts to incorporate empirical methods into philosophy are not new. In fact, many great philosophers of old were inspired by and sought to emulate the success of science pretty much since the beginning of modern science itself (Voegelin 1948). To mention just a few obvious examples: Thomas Hobbes' *Leviathan* (1651) introduced concepts from the new science of the seventeenth century, such as force and endeavor, into social and political philosophy, Baruch Spinoza's *Ethics* (1677) incorporated the geometrical method into metaphysics and moral philosophy, and David Hume's *A Treatise on Human Nature* (1739–1740) was an "attempt to introduce the experimental method of reasoning into moral subjects."[14] To this list, Catherine Wilson (2019, 55) adds Democritus, Aristotle, Epicurus, Locke, Leibniz, Kant, Nietzsche, and Mach: all of whom "took a direct interest in medicine, natural history, cosmology, evolutionary theory, anthropology, the visual system, and physics and drew philosophical inspiration from these fields." I suspect that no one would find these great philosophers guilty of "scientific expansionism" or "scientific imperialism." I also suspect that no one would complain that these giants of philosophy were not doing philosophy.[15] They were not territorial about philosophy. By incorporating methods from the successful sciences of their time, these great philosophers have produced some of the most enduring works of philosophy. Rather than fear and resist the introduction of scientific ideas, methods, and practices into philosophy, then, we should follow their example and embrace it. We might not have another Aristotle, Spinoza, or Kant, of course, but we might still have some original work produced, nonetheless.

NOTES

1. For example, see my exchanges with Christopher Brown and Bernard Wills in the *Social Epistemology Review and Reply Collective*, Mizrahi (2017b), (2018a), (2018b), and (2018c). In these papers, I distinguish between *Strong Scientism* and *Weak Scientism*. The former is the view that scientific knowledge is the *only* knowledge we have, whereas the latter is the view that scientific knowledge is the *best* knowledge we have. The argument for Weak Scientism is that scientific knowledge is quantitatively better (in terms of research output and research impact) and qualitatively better (in terms of explanatory, instrumental, and predictive success) than non-scientific knowledge. See chapter 1 of this volume for a more detailed discussion.

2. For further discussion of the "scientism" charge, see chapter 1 of this volume.

3. On the so-called method of cases and the use of intuitions as evidence in academic philosophy, see Mizrahi (2014), (2015a), (2015b), and (2021b).

4. It is important to note that these hypotheses are not meant to be exhaustive. There may be other reasons why academic philosophers might find scientism threatening to academic philosophy. For further discussion, see chapters 3, 5, and 6 of this volume.

5. See also Wiebe et al. (2005) for more details on sentiment analysis in NLP and the MPQA Subjectivity Lexicon.

6. Contrary to what Jeroen de Ridder (2019, 10) seems to think, "explain" does not necessarily mean a *causal* explanation here. "The R-squared (R^2) measures the explanatory or predictive power of a regression model. It is a GOODNESS-OF-FIT MEASURE, indicating how well the linear regression equation fits the data" (Lewis-Beck 2004, 983). Accordingly, when we say that 93 percent of the variation in the percentage of Philosophy and Religious Studies bachelor's degrees is explained by the percentage of STEM bachelor's degrees by year, "this explanation may be more 'statistical' than 'causal'" (Lewis-Beck 2004, 983). It could be that STEM bachelor's degrees conferred over the years help *predict* variations in Philosophy and Religious Studies bachelor's degrees but do not really explain them in a *causal* sense. In other words, it may be "that X [for example, STEM bachelor's degrees by year] merely 'accounts for' so much variation in Y [for example, Philosophy and Religious Studies bachelor's degrees]" (Lewis-Beck 2004, 983).

7. For additional examples of the application of methods from data science, such as text mining, corpus analysis, and data visualization, to philosophical research, see Ashton and Mizrahi (2018a) and Ashton and Mizrahi (2018b).

8. See also Knobe and Nichols (2017).

9. Again, contrary to what Jeroen de Ridder (2019, 10) seems to think, "explain" does not necessarily mean a *causal* explanation here. "The R-squared (R^2) measures the explanatory or predictive power of a regression model. It is a GOODNESS-OF-FIT MEASURE, indicating how well the linear regression equation fits the data" (Lewis-Beck 2004, 983). Accordingly, when we say that 95 percent of the variation in the number of "scientism" publications is explained by the number of "experimental philosophy" publications by year, "this explanation may be more 'statistical'

than 'causal'" (Lewis-Beck 2004, 983). It could be that "experimental philosophy" articles published over the years help *predict* variations in "scientism" publications but do not really explain them in a *causal* sense. In other words, it may be "that *X* [for example, 'experimental philosophy' publications by year] merely 'accounts for' so much variation in *Y* [for example, 'scientism' publications]" (Lewis-Beck 2004, 983).

10. For a discussion of Sorell's "scientism" charge against experimental philosophy, see chapter 9 of this volume.

11. For further discussion, see Mizrahi (2017a).

12. See chapter 9 of this volume for a detailed discussion of Sorell's (2018) argument.

13. See chapter 9 of this volume for further discussion.

14. See chapter 1 of this volume for further discussion.

15. Opponents of scientism and critics of experimental philosophy often complain that empirical work is "not philosophy" (Jenkins 2014) or that it is not "philosophically significant" (Kauppinen 2007). Cf. Knobe (2007) and O'Neill and Machery (2014).

REFERENCES

Ashton, Zoe and Mizrahi, Moti. 2018a. "Intuition Talk is Not Methodologically Cheap: Empirically Testing the 'Received Wisdom' About Armchair Philosophy." *Erkenntnis* 83 (3): 595–612.

Ashton, Zoe and Mizrahi, Moti. 2018b. "Show Me the Argument: Empirically Testing the Armchair Philosophy Picture." *Metaphilosophy* 49 (1–2): 58–70.

Bilgrami, Akeel. 2014. "The Humanities in the English-Speaking West." *Social Scientist* 42 (9–10): 43–47.

Blease, Charlotte. 2017. "Philosophy's Territorialism: Scientists Can Talk About Values Too." *Philosophy, Psychiatry & Psychology* 24 (3): 231–234.

Browne, Kayali Tamara. 2017. "A Role for Philosophers, Sociologists and Bioethicists in Revising the DSM: A Philosophical Case Conference." *Philosophy, Psychiatry & Psychology* 24 (3): 187–201.

Buckwalter, Wesley and Turri, John. 2018. "Moderate Scientism in Philosophy." In J. de Ridder, R. Peels, and R. van Woudenberg (eds.). *Scientism: Prospects and Problems* (pp. 280–300). New York: Oxford University Press.

Chalmers, David. 2015. "Why Isn't There More Progress in Philosophy? *Philosophy* 90 (1): 3–31.

Churchland, Patricia, S. 2011. *Braintrust: What Neuroscience Tells Us about Morality*. Princeton, NJ: Princeton University Press.

De Ridder, Jeroen. 2019. "Against Empirical-ish Philosophy: Reply to Mizrahi." *Social Epistemology Review and Reply Collective* 8 (12): 8–12.

Douglas, Heather E. 2009. *Science, Policy, and the Value-Free Ideal*. Pittsburgh, PA: University of Pittsburgh Press.

Dupré, John. 2001. *Human Nature and the Limits of Science*. Oxford: Clarendon Press.

Foot, Philippa. 1958. "Moral Arguments." *Mind* 67 (268): 502–513.

Haack, Susan. 2017. "The Real Question: Can Philosophy Be Saved?" *Free Inquiry* 37 (6): 40–43.

Haug, Matthew, C. 2014. "Introduction--Debates about Methods: From Linguistic Philosophy to Philosophical Naturalism." In M. C. Haug (ed.), *The Armchair or the Laboratory?* (pp. 1–26). London: Routledge.

Jenkins, Katharine. 2014. "'That's not philosophy': Feminism, Academia, and the Double Bind." *Journal of Gender Studies* 23 (3): 262–274.

Kauppinen, Antti. 2007. "The Rise and Fall of Experimental Philosophy." *Philosophical Explorations* 10 (2): 95–118.

Kitcher, Philip. 2017. "The Trouble with Scientism: Why History and the Humanities are also a Form of Knowledge." In *Science Unlimited? The Challenges of Scientism*, edited by M. Boudry and M. Pigliucci, 109–120. Chicago: The University of Chicago Press.

Knobe, Joshua. 2007. "Experimental Philosophy and Philosophical Significance." *Philosophical Explorations* 10 (2): 119–121.

Knobe, Joshua. 2015. "Philosophers are Doing Something Different Now: Quantitative Data." *Cognition* 135 (2015): 36–38.

Knobe, Joshua and Nichols, Shaun. 2008. "An Experimental Philosophy Manifesto." In J. Knobe and S. Nichols (eds.), *Experimental Philosophy* (pp. 3–14). New York: Oxford University Press.

Knobe, Joshua and Nichols, Shaun. 2017. "Experimental Philosophy." In E. N. Zalta (ed.), *The Stanford Encyclopedia of Philosophy* (Winter 2017 Edition). https://plato.stanford.edu/archives/win2017/entries/experimental-philosophy/.

Kuhn, Thomas S. 2000. "Afterwords." In *The Road Since Structure: Philosophical Essays, 1970–1993, with an Autobiographical Interview*, edited by J. Conant and J. Haugeland, 224–252. Chicago: The University of Chicago Press.

Kumar, Keerthi H. M. and Harish, B. S. 2020. "A New Feature Selection Method for Sentiment Analysis in Short Text." *Journal of Intelligent Systems* 29 (1): 1122–1134.

Lewis-Beck, Michael S. 2004. "R-Squared." In *The SAGE Encyclopedia of Social Science Research Methods*, Volume 3, edited by Michael S. Lewis-Beck, Alan Bryman, and Tim Futing Liao, 983–984. Thousand Oaks, CA: SAGE Publications, Inc.

Liu, Bing. 2017. "Many Facets of Sentiment Analysis." In *A Practical Guide to Sentiment Analysis*, Socio-Affective Computing, Volume 5, edited by Erik Cambria, Dipankar Das, Sivaji Bandyopadhyay, and Antonio Feraco, 11–39. Cham, Switzerland: Springer.

Mizrahi, Moti. 2014. "Does the Method of Cases Rest on a Mistake?" *Review of Philosophy and Psychology* 5 (2): 183–197.

Mizrahi, Moti. 2015a. "Don't Believe the Hype: Why Should Philosophical Theories Yield to Intuitions?" *Teorema: International Journal of Philosophy* 34 (3): 141–158.

Mizrahi, Moti. 2015b. "Three Arguments Against the Expertise Defense." *Metaphilosophy* 46 (1): 52–64.

Mizrahi, Moti. 2017a. "What's So Bad about Scientism?" *Social Epistemology* 31 (4): 351–367.

Mizrahi, Moti. 2017b. "In Defense of *Weak Scientism*: A Reply to Brown." *Social Epistemology Review and Reply Collective* 6 (11): 9–22.

Mizrahi, Moti. 2018a. "More in Defense of Weak Scientism: Another Reply to Brown." *Social Epistemology Review and Reply Collective* 7 (4): 7–25.

Mizrahi, Moti. 2018b. "Weak Scientism Defended Once More." *Social Epistemology Review and Reply Collective* 7 (6): 41–50.

Mizrahi, Moti. 2018c. "Why Scientific Knowledge Is Still the Best." *Social Epistemology Review and Reply Collective* 7 (9): 18–32.

Mizrahi, Moti. 2019. "What Isn't Obvious about 'Obvious': A Data-Driven Approach to Philosophy of Logic." In A. Aberdein and M. Inglis (eds.). *Advances in Experimental Philosophy of Logic and Mathematics* (pp. 201–224). London: Bloomsbury.

Mizrahi, Moti. 2021a. "Conceptions of Scientific Progress in Scientific Practice: An Empirical Study." *Synthese*. 199 (1-2): 2375–2394.

Mizrahi, Moti. 2021b. Your Appeals to Intuition Have No Power Here! *Axiomathes*. doi: 10.1007/s10516-021-09560-9.

Mizrahi, Moti. 2022. "What is the Basic Unit of Scientific Progress? A Quantitative, Corpus-Based Study." *Journal for General Philosophy of Science*. DOI: https://doi.org/10.1007/s10838-021-09576-0.

Nagel, Jennifer. 2012. "Intuitions and Experiments: A Defense of the Case Method in Epistemology." *Philosophy and Phenomenological Research* 85 (3): 495–527.

Nussbaum, Martha, C. 2010. *Not For Profit: Why Democracy Needs the Humanities*. Princeton, NJ: Princeton University Press.

Oddie, Graham. 2006. "A Refutation of Peircean Idealism." In C. Cheyne and J. Worrall (eds.), *Rationality and Reality: Conversations with Alan Musgrave* (pp. 255–262). Dordrecht: Springer.

O'Neill, Elizabeth and Machery, Edouard. 2014. "Experimental Philosophy: What Is It Good For?" In E. Machery and E. O'Neill (eds.), *Current Controversies in Experimental Philosophy* (pp. vii–xxix). New York: Routledge.

Pigliucci, Massimo. 2018. "The Problem with Scientism." *Blog of the APA*, January 25, 2018. https://blog.apaonline.org/2018/01/25/the-problem-with-scientism/.

Putnam, Hilary. 2002. *The Collapse of the Fact/Value Dichotomy and Other Essays*. Cambridge, MA: Harvard University Press.

Schmidt, Benjamin. 2018. "The Humanities are in Crisis." *The Atlantic*, August 23, 2018. https://www.theatlantic.com/ideas/archive/2018/08/the-humanities-face-a-crisisof-confidence/567565/.

Scruton, Roger. 2014. *The Soul of the World*. Princeton, NJ: Princeton University Press.

Shapiro, Rees, T. 2017. "For Philosophy Majors, the Question After Graduation Is: What Next?" *The Washington Post*, June 20, 2017. https://www.washingtonpost.com/local/education/for-philosophy-majors-the-question-after-graduation-is-what

-next/2017/06/20/aa7fae2a-46f0-11e7-98cd-af64b4fe2dfc_story.html?utm_term=
.f72d3cac2d58.

Sorell, Tom. 2017. "Scientism (and Other Problems) in Experimental Philosophy." In M. Boudry and M. Pigliucci (eds.), *Science Unlimited? The Challenge of Scientism* (pp. 263–282). Chicago: The University of Chicago Press.

Sorell, Tom. 2018. "Experimental Philosophy and the History of Philosophy." *British Journal for the History of Philosophy* 26 (5): 829–849.

Stenmark, Mikael. 1997. "What is Scientism?" *Religious Studies* 33 (1): 15–31.

Van De Poel, Ibo. 2020. "Should Philosophers Begin to Employ New Methods if They Want to Become More Societally Relevant?" In Julia Hermann, Jeroen Hopster, Wouter Kalf, and Michael Klenk (eds.), *Philosophy in the Age of Science? Inquiries into Philosophical Progress, Method, and Societal Relevance* (pp. 231–244). London: Rowman & Littlefield.

Van Inwagen, Peter. 2004. "Freedom to Break the Laws." *Midwest Studies in Philosophy* 28 (1): 334–350.

Voegelin, Eric. 1948. "The Origins of Scientism." *Social Research* 15 (4): 462–494.

Wiebe, Janyce, Wilson, Theresa, and Cardie, Claire. 2005. "Annotating Expressions of Opinions and Emotions in Language." *Language Resources and Evaluation* 39 (2): 165–210.

Wilson, Catherine. 2019. "Regarding Scientism and the Soul of Philosophy." *Social Epistemology Review and Reply Collective* 8 (11): 55–58.

Wilson, Theresa, Wiebe, Janyce, and Hoffman, Paul. 2005. "Recognizing Contextual Polarity in Phrase-Level Sentiment Analysis." *Proceedings of the Conference on Human Language, Technology, and Empirical Methods in Natural Language Processing*. Association for Computational Linguistics. doi: 10.3115/1220575.1220619.

Chapter 3

The Supposed Specter of Scientism

Amanda Bryant

This chapter considers the assumptions required to make scientisms of different forms genuinely threatening to philosophers, where a genuine threat would consist of a concrete risk to their statuses, the value of their teaching and research, their livelihoods, their preferred research methods, or the health of the discipline. This chapter will argue that strong and weak forms of scientism alike require substantive assumptions to make them threatening in those regards. In particular, they require sometimes heavy-handed circumscriptions of philosophy and science, as well as their epistemic credentials and achievements, methods, and subject matters. They also require restrictive pronouncements upon the epistemic and non-epistemic goods that are valuable, worth promoting in academic contexts, and relevant to disciplinary health. My aim in this chapter is not to prove those assumptions false but rather to make them explicit and to emphasize their frequent strength and contentiousness.

1. IMAGES OF SCIENTISM

There is a kind of arrogantly dismissive attitude toward non-scientific fields of inquiry and ways of knowing—often associated with publicly prominent physicists and so-called new atheists—which is sometimes called "scientism" (the kind described by, e.g., Pigliucci 2018). There is also a kind of admiringly respectful attitude toward science, paired with an optimism regarding its import for traditionally non-scientific questions—often associated with avowedly naturalistic philosophers—which is sometimes called "scientism" (the kind espoused by, e.g., Ladyman et al. 2007). Some philosophers pick fights about what the term *really* means or what *really* falls under its rubric.

47

Yet there is no use warring over the label and its essential connotations, since terms of art such as this can mean what we want them to.[1]

However, the history of a term can make us less terminologically flexible than we might be ideally. Defenders of scientism face an uphill battle to convince others of the palatability of at least some scientisms. Part of the difficulty stems from the history of the label, in which the term was inherently pejorative (Dupré 1993; Haack 2003; Sorell 1991). Notwithstanding the term's well-known reappropriation by naturalists such as Rosenberg (2011) and Ladyman et al. (2007), defenders of scientism are apparently still working to unsaddle the term of its baggage.

For instance, Ladyman (2018) proposes to give scientism *a humane face*, which suggests that its usual face is generally thought to be inhumane. Buckwater and Turri (2018) defend a *moderate* form of scientism, which suggests that the usual form is generally thought to be immoderate. Mizrahi (2017) asks *what's so bad* about scientism and defends what he calls *Weak Scientism*, which suggests that scientism is generally thought to be both bad and unduly strong. These authors think that scientism need not be inherently vicious or threatening, but rather that philosophy can benefit from adopting some appropriately temperate form of it. If these defenses of "nice" scientism are any indication, defenders of scientism are working to overcome a rather mean-looking image.

So there is a clear sociological narrative according to which scientism is generally viewed by philosophers as (echoing Haack 2017, 41) a specter, to be regarded with fear or hostility. Mizrahi (2019) even sets out to empirically test possible explanations as to why many philosophers find scientism threatening. But even if some philosophers are threatened by it—as evidenced by the occasional "ferociousness" with which they respond to it (Mizrahi 2019, 1)—there is still a question as to the prevalence of the sense of threat. In a well-known study, Bourget and Chalmers showed that "philosophers have substantially inaccurate sociological beliefs about the views of their peers" (2014, 489). This shows the importance of substantiating our sociological narratives. We ought to empirically confirm that the running sociological narrative represents philosophical sentiment.

However, without introducing some operative definition of scientism, the question of *how philosophers really feel about scientism* is too ambiguous to constitute a well-defined research question. As we have seen, in the broadest strokes, there are at least two competing images of scientism: the mean one and the comparatively nice one. To complicate matters further, attempts to generate a specific and contentful formulation of scientism have produced vastly many substantively distinct theses, both of the inherently negative variety (Haack 2003, 2012, 2017; Pigliucci 2010; Sorell 1991) and of nonnegative varieties (Buckwalter and Turri 2018; Ladyman et al. 2007; Mizrahi 2017; Rosenberg 2011). This has, in turn, resulted in the need to catalog, compare, and taxonomize the various formulations (Hietanen et al. 2020; Peels 2018; Stenmark 2018). So at this complicated juncture in the dialectic, it is clear that there is no one thing, *scientism*, about which we can gauge overall philosophical sentiment. Rather, there is a range of scientisms, our attitudes toward which require much more directed and detailed sociological investigation. So what we need in order to test our sociological narrative is unambiguous hypotheses; then, since such hypotheses would concern conscious individual feelings, the natural way to proceed (as de Ridder 2019 and Wilson 2019 point out) would be via survey and interview.

Setting aside the sociological question of how philosophers actually feel about various scientisms, I propose to consider whether they *should* feel threatened by any forms of scientism and, if so, which ones and why. For the purposes of this chapter, for philosophers to "feel threatened" by a form of scientism is for them to believe that its truth—or perceived truth—would harm, disrupt, or undermine their professional standing or certain other things they value. Inter alia, philosophers feeling threatened by scientism could involve their believing it poses some concrete risk to:

- their prestige or status
- the value of their teaching and research
- their livelihood
- the fruitfulness or continuation of their preferred research methods
- the health of the discipline

Given this characterization of what it means for philosophers to "feel threatened" by a form of scientism, it follows that for it to be the case they *should* feel so threatened would be for them to be *correct* in their belief that the truth or perceived truth of the relevant form of scientism could potentially harm, disrupt, or undermine the sorts of things I have listed. What I am interested in is what sorts of epistemological, methodological, and value-theoretic assumptions would warrant such a belief. So, ultimately, my question is: which, if any, forms of scientism are threatening, on which assumptions?

Section 2 will distinguish a number of scientisms. Section 3 will spell out the sorts of assumptions required to make strong scientisms—which are the most prima facie threatening scientisms—plausibly true and plausibly threatening to philosophers in the ways I have spelled out. Section 4 will do the same for Mizrahi's "Weak Scientism," and section 5 will conclude.

2. SCIENTISMS

To answer the question of which forms of scientism are threatening under which assumptions, I must first distinguish a variety of scientisms. Space limitations allow me to consider just a smattering of available positions, but I hope it to be adequately representative. For a fuller taxonomy of scientisms and their logical relations to one another, see Peels (2018).

First, consider the inherently negative variety, which builds in elements such as unwarranted, exaggerated, excessive, or uncritical attitudes. For instance, take Susan Haack's characterization of scientism as "inappropriate, uncritical deference to the sciences" (2017, 41). The question of whether this sort of scientism could potentially harm, disrupt, or undermine philosophy is not particularly interesting. In fact, the answer is virtually trivial. That's because it doesn't take a particularly strong or controversial view of what's healthy or unhealthy for the discipline to think that *any* unwarranted, exaggerated, excessive, or uncritical philosophical attitudes would, if sufficiently prevalent, be bad for it. Unwarranted, exaggerated, and excessive attitudes are inherently inappropriate, so their prevalence in philosophy would be intrinsically bad. At any rate, insofar as I see questions regarding the advisability, worth, and epistemic virtuousness or viciousness (Kidd 2018) of scientism as both live and interesting, I believe we should prefer neutral definitions.[2,3] With that said, I set the inherently negative definitions aside.

As far as neutral definitions go, the category is remarkably heterogeneous. Consider the following theses, in which slashes indicate alternate formulations, and which vary in substance (epistemological, methodological, and disciplinary) and strength (strong, moderate, and weak):

Strong Epistemological Scientism: Science is the *only/only good* source of certain epistemic goods (knowledge, justified belief, evidence).

Moderate Epistemological Scientism: Science is the *best* source of certain epistemic goods.

Weak Epistemological Scientism: Science is a *comparatively excellent* source of certain epistemic goods.

Strong Methodological Scientism: Given their epistemic aims, inquirers/philosophers should *only* use the methods of science.

Moderate Methodological Scientism: Given their epistemic aims, inquirers/philosophers should *primarily* use the methods of science.

Weak Methodological Scientism: Given their epistemic aims, inquirers/philosophers should *to some extent* use the methods of science.

Strong Disciplinary Scientism: Science will/should subsume/replace *all* other forms of inquiry.

Moderate Disciplinary Scientism: Science will/should subsume/replace *most* other forms of inquiry.

Weak Disciplinary Scientism: Science will/should subsume/replace *some* other forms of inquiry.

The list is neither exhaustive nor authoritative. Other variations are certainly possible. For instance, methodological naturalism might be formulated non-prescriptively as a value-theoretic claim stating that scientific methods are superior to all others.[4] Of course, such a claim might be thought to lead naturally to the prescriptive claim that inquirers *should* use those methods.

What I have called "disciplinary" scientism can be interpreted as a form of "scientific imperialism," according to which the boundaries of the discipline or institution of science will or should extend outward to enfold erstwhile independent disciplines such as philosophy. Such scientisms claim neither that science has any special role in securing epistemic goods (epistemological scientism) nor that philosophy should use scientific methods (methodological scientism), but rather that philosophy will or should become part of science or be eliminated in its favor, depending on the formulation. The precise content of the thesis needs to be specified. For instance, one possible disciplinary thesis could prescribe that faculties of arts be disbanded and that departments such as philosophy be enfolded under faculties of science. An alternate thesis could prescribe that philosophers be fired and their jobs given to scientists. There is a lot of room for variation.

Epistemological, methodological, and disciplinary varieties of scientism should not be conflated. Epistemological varieties concern the capacity of science to secure epistemic goods; methodological varieties concern the integration of scientific methods into philosophy or other areas of inquiry; disciplinary varieties concern disciplinary boundaries. These different sorts of thesis might interrelate in interesting ways. For instance, the idea that science has a monopoly on knowledge (strong epistemological scientism) could motivate the claim that science should replace all other forms of inquiry (strong disciplinary scientism). Notwithstanding such possible connections, the distinctions among these scientisms should be kept clearly in mind.

Whether all of the above theses count as bona fide forms of scientism is an open question. However, as I indicated at the outset, I take it to be a merely terminological one, to which there is no definitive or interesting answer. Whether all candidate formulations of scientism could be unified under a single overarching thesis is another open question. However, I believe the prospects for unification to be slim, given the heterogeneity of available conceptions.

There are still further distinctions to be drawn, in addition to the ones already highlighted. For instance, while Rosenberg claims that "Thoroughgoing scientism leaves no room for normative values" (2020, 50), some scientisms are themselves thoroughly normative, since some make methodological prescriptions, and since some (I would argue most) are premised on the epistemic or pragmatic value of science. Other formulations of scientism are modal, such as those that claim that only science *can* address certain kinds of questions. Others still are descriptive, such as disciplinary theses claiming that science *will* subsume or replace other forms of inquiry. Moreover, we can distinguish scientisms not only in terms of strength but also in terms of scale. For instance: a *global* form of epistemological scientism might say that science is the best source of evidence regarding *any question whatsoever*. A comparatively *local* form might say that science is the best source of evidence regarding *certain sorts of questions* (such as metaphysical or moral questions). So scientisms have many distinguishing features and differ in a variety of interesting and important ways.

These sorts of distinctions ought always to inform our discussions of scientism. Some philosophical terms are so thoroughly contested and variably defined that we cannot hope to have a productive conversation about them without disambiguating clearly at the outset. Scientism is one of those terms. Failing to disambiguate leads to ignored distinctions, misconstruals, and cross-purposes, as well as obscured lines between scientism and a constellation of nearby epistemological and methodological positions and practices, such as empiricism, physicalism, realism, naturalism, and interdisciplinarity.[5] So I hereby plead, if we're going to talk about "scientism," let's always start with a definition.[6]

3. MAKING SPECTERS OF SCIENTISMS

The question at hand is: which assumptions warrant a sense of threat among philosophers in response to which forms of scientism? In this section, I will identify a number of assumptions required to make scientisms plausibly true and plausibly threatening to philosophers' statuses, livelihoods, preferred research methods, and so forth. While the assumptions are compatible with scientism, they are not entailed by it and therefore not necessarily held by its proponents. Rather, all of the assumptions must be *appended* to scientism and held by individual philosophers who find scientism threatening, in order for their sense of threat to be warranted. I will focus on the most prima facie threatening forms of scientism—the strong scientisms—since, given their strong formulations, it would not be surprising if they turned out to be most threatening in the ways I have outlined. I will not aim to defeat the assumptions here but rather to make them explicit and to emphasize their frequent strength and contentiousness.

3.1. Strong Epistemological Scientism

First, take the strong epistemological view that science is the *only* source of certain epistemic goods like knowledge. What are the implications of this particular scientism for philosophers? It might be thought to imply that science can get, let's say, knowledge *while philosophy can't.*[7] But this is only implied if one takes the following assumption for granted:

(1) *Demarcation*: Philosophy isn't science.

It might initially appear that this assumption is entailed by scientism, but it need not be. Proponents of scientism can recognize a continuum in which philosophy and science aren't cleanly demarcated, but bleed into one another, and yet think that the philosophical side of the continuum stretches too far from the scientific side. On such a view, it just isn't true that science *and not philosophy* can get knowledge. Now, it is trivially true from an institutional perspective that philosophy isn't science—philosophers have their own departments, and scientists have theirs. However, it is not as obviously true from a philosophical standpoint. That's because we lack adequate, uncontroversial definitions of science and philosophy in virtue of which we can definitively distinguish them. The demarcation problem in the philosophy of science is famously intractable. Moreover, we lack a good answer to parallel questions about the distinctive nature of philosophy and the delineation of its borders. So *Demarcation* lacks an adequate philosophical foundation.

You might reasonably suggest that even if we cannot resolve these demarcation issues by supplying necessary and sufficient conditions, it is common sense that philosophy differs from science in certain important and identifiable respects, some of which are undoubtedly relevant to knowledge. After all, many scientists spend their time using highly sophisticated and expensive equipment to test predictions that have (given certain auxiliary hypotheses) specific observable consequences, while philosophers . . . don't. So, you might think that in order to derive the conclusion that philosophy can't get knowledge from the claim that only science can, you need not solve the demarcation problem by identifying all the essential differences between philosophy and science; you need only assume that science and philosophy differ *in epistemically relevant ways*.

(2) *Epistemic Distinctness*: Science and philosophy are distinct in ways relevant to the acquisition of the epistemic goods on which science has a monopoly.

For example, you might say that science and philosophy differ qua sources of knowledge. Philosophy must lack certain qualities that science possesses, in virtue of which science is knowledge-conducive, or else philosophy must possess certain knowledge-compromising qualities that science lacks. The epistemology of science and the epistemology of philosophy must diverge in some epistemically difference-making respect or respects, which the defender of strong epistemological scientism must identify.

Suppose we were to grant that "only science gets knowledge" implies that philosophy doesn't. Strong epistemological scientism would still not obviously threaten philosophers' statuses, the value of their teaching and research, or their livelihoods. On the one hand, we would have to grant that philosophers' attempted contributions to knowledge universally fail. On the other hand, that would not mean that it is impossible for philosophy to produce other epistemic goods. For instance, if the particular scientism at issue says that science is the only source of knowledge, it leaves open the possibility that philosophy might be a source of epistemic goods such as justified belief, evidence, understanding, and so forth. In order for it to follow that philosophy is epistemically worthless, we would have to assume:

(3) *Epistemic Value Monism*: The epistemic goods on which science has a monopoly are the only genuine epistemic goods.

This precludes justified belief, evidence, understanding, or other putative epistemic goods from counting as bona fide epistemic goods, that is, as epistemically valuable in their own right. Moreover, in order for strong epistemological scientism to undermine philosophers' statuses, the value of their teaching and research, or their place in the university, we would have to assume something like:

(4) *Professional Value (Restrictive)*: For philosophers to be considered serious academics, for their teaching and research to be valuable, or for them to have a proper place as faculty members in universities require that they promote those epistemic goods on which science has a monopoly.

In other words, promoting other sorts of good is insufficient for philosophers and their work to be valuable and respected. But if promoting other goods is insufficient to make philosophy valuable, that must mean that no other goods are valuable in and of themselves. If so, then *Professional Value (Restrictive)* entails that the *epistemic* goods on which science has a monopoly are the only *goods simpliciter* worth recognizing, preserving, or promoting in academic contexts. To accept this, we would have to deny that aesthetic or pragmatic goods are genuinely valuable in academic contexts—or valuable enough to be worth pursuing for their own sakes. So strong epistemological scientism threatens philosophers' statuses, work, and livelihoods only holding fixed an extremely restrictive view regarding value in academic contexts.

What about the implications of strong epistemological scientism for philosophers' preferred research methods? Well, if science sometimes gets knowledge and philosophy never does, then whatever methods science employs are sometimes knowledge-conducive, and whatever methods philosophy employs aren't ever. For this story to be plausible, we have to assume:

(5) *Methodological Distinctness*: We can distinguish distinctively "scientific" from distinctively "philosophical" methods.[8]

In other words, beyond saying that a method happens to be implemented in philosophical or scientific practice, we can point to certain methods and say *those are squarely scientific methods*, and we can point to other methods and say *those are squarely philosophical methods*. The question is how—and on what basis—to decide which methods are which. A historical who-used-it-first approach to the question would risk giving a historically contingent answer to a deeper philosophical question (not to mention that philosophy would have an unfair advantage, since it predates the advent of modern science).

How else might we distinguish distinctively philosophical from distinctively scientific methods? Sometimes the tendency is to believe that philosophical method is essentially a priori (a belief that Mizrahi 2019, for example, attributes to many philosophers), while scientific method is essentially a posteriori. There is a morsel of truth here, since, as I said above, scientists often do test hypotheses empirically, while philosophers often don't (and often can't). However, we must also take care not to rely on imprecise or simplistic caricatures. The ambiguity of the term "a priori" (see Field 2005) and, correspondingly, of the term "a posteriori" complicates matters. If a method's apriority requires that it not involve experience in any respect, then arguably even "traditional" philosophical reflection fails to be a priori. That is because it is often beholden to explanatory demands that invoke judgments of plausibility, unifying power, and likelihood *relative to a background of experientially based beliefs* (Bryant 2020). Moreover, Kidd points out that "philosophy is very methodologically pluralistic and has welcomed and, indeed, often introduced empirical methods" (2019, 53). It isn't clear why we should think that, in so doing, philosophical practice has diverged from *essentially* philosophical method. So it is not obvious that philosophical method must be or always is a priori.

At the same time, as Chakravartty (2013) points out, science isn't purely a posteriori, either. He explains, "not all sciences actually make novel predictions (evolutionary biology), or employ experiments (string theory), or

are successful in manipulating things (cosmology)" (2013, 34). To state the
point more positively, scientists implement a range of methods that might be
considered a priori, including thought experiments (Einstein's train); comput-
ability theory, modal logic, and category theory (used in computer science);
mathematics (ubiquitous in science); and even armchair speculation (string
theory).[9] Given these sorts of considerations, I believe that equivocating
between "scientific" and "empirical" is ill-advised. At any rate, distinguishing distinctively philosophical and scientific methods by appealing to at least
some of the characterizations of the a priori-a posteriori distinction is clumsy
and misleading. If we are to find some plausible basis for distinguishing them,
it will have to be a good deal subtler.

Supposing that philosophy and science do have their own distinctive
methods, in order for it to be plausible that scientific method is sometimes
knowledge-conducive, while philosophical method never is, we would need
an account of the differences between them that explain their differential success. One possibility is that scientific method has distinctive components that
are necessary for knowledge, which philosophical method lacks. To capture
this thought, we would need the following assumptions:

(6) *Scientific Indispensability*: No combination of distinctively philosophical
    methods is sufficient to produce the epistemic goods on which science
    has a monopoly: producing them requires using some distinctively scien-
    tific methods.
(7) *Philosophical Failure*: Philosophers either never use or never *success-
    fully* use the distinctively scientific methods implicated in the success of
    science with respect to its proprietary epistemic goods.

In other words, scientists can get knowledge and philosophers can't *because*
scientists do certain distinctively scientific things that philosophers don't.
Scientific Indispensability implies that if certain reflective methods, such as
the use of thought experiments, count as distinctively philosophical, their
implementation is insufficient for knowledge even in some scientific contexts. For instance, it implies that Einstein's armchair speculations did not
become objects of knowledge until they were confirmed empirically. This
sort of view is defensible but clearly contentious.

By contrast, *Philosophical Failure* is less defensible, since it suggests
that the philosophers who implement scientific methods (of which there are
many) just happen never to implement the knowledge-conducive ones, or
else always screw up when they do. Since philosophers use a wide variety
of scientific methods, it would be spectacularly unlucky (not to mention
incredibly bad news for science) if none of those methods are knowledge-
conducive. Moreover, while it's clear that philosophers sometimes attempt

to use scientific methods that they haven't been adequately trained to use and fail (poorly designed statistical surveys come to mind), it's wildly implausible that no philosopher who uses scientific methods ever does so correctly. *Philosophical Failure* also implies that if science and philosophy share certain methodological features in common—such as evidential or rational standards, underlying logic or inferential patterns, or criteria for theory selection—then, regardless of whether those features count as distinctively philosophical or distinctively scientific, insofar as philosophers implement them correctly, they can never be independently adequate for knowledge. At any rate, a defender of strong epistemological scientism who explains the differential epistemic standing of science and philosophy by appealing to *Scientific Indispensability* and *Philosophical Failure* will have to identify the distinctively scientific ingredients that are necessary for the relevant epistemic successes.

In addition to it being possible, according to strong epistemological scientism, that philosophy fails to produce knowledge because its practice excludes knowledge-conducive methods, it is also possible that its epistemic failures are due to its reliance on falsity-conducive or justification-undermining methods. If so, then we need the following assumptions:

(8) *Unique Culpability*: All methods implicated in the failure of philosophy vis-à-vis the relevant epistemic goods are distinctively philosophical.
(9) *No Contamination*: When science successfully produces its proprietary epistemic goods, its practice includes none of the methods implicated in the failure of philosophy vis-à-vis those goods.

In other words, philosophers always fail to get knowledge while scientists at least sometimes don't *because* philosophers do certain distinctively philosophical things that scientists at least sometimes don't. For example, if it is the speculative aspects of philosophical method that doom philosophy's epistemic prospects, then epistemically successful science must implement no such speculative methods.[10] Note how much heavy-handed methodological circumscription these assumptions require. At any rate, if the defender of strong epistemological scientism opts for this explanation of differential success, then she must identify the failure-conducive features of philosophical method.

One final question before we can move on from strong epistemological scientism: Does the view that science is the only source of epistemic goods like knowledge, if true or if broadly believed to be true, threaten the health of the discipline of philosophy? It does only on the following assumption:

(10) *Disciplinary Health (Restrictive)*: The health of the discipline of philosophy requires that philosophy produce the epistemic goods on which science has a monopoly.

For example, we would have to assume that the health of the discipline requires that philosophy produce knowledge. If we were to allow, *contra* (3) above, *Epistemic Value Monism*, that there are other genuine epistemic goods, it would not be sufficient for disciplinary health that philosophy produces them. Likewise, if we were to allow, *contra* (4) above, *Professional Value (Restrictive)*, that there are non-epistemic goods worth pursuing in academic contexts, it would not be sufficient for disciplinary health that philosophy promotes them. On the contrary, *Disciplinary Health (Restrictive)* makes producing the epistemic goods on which science has a monopoly a *necessary* condition of disciplinary health and implies, as such, that failure to produce such goods immediately compromises disciplinary health.

To sum things up so far, I have been cataloging the assumptions required to make strong epistemological scientism, according to which science is the only source of certain epistemic goods such as knowledge, threatening to philosophers. I suggested that the view threatens philosophers' statuses, the value of their teaching and research, as well as their livelihoods only if we draw heavy-handed taxonomic and epistemological distinctions between philosophy and science, and only if we recognize a narrow range of epistemic and non-epistemic goods worth preserving and promoting in academic contexts. Moreover, strong epistemological scientism impugns philosophical method only if we assume there are distinctively philosophical research methods that differ from scientific ones in identifiable ways, which difference adequately explains their differential epistemic success. Finally, the truth of strong epistemological scientism implies compromised disciplinary health only if we assume that to be healthy, philosophy must produce precisely those epistemic goods that, according to strong epistemological scientism, it cannot possibly produce.

I have not argued that all of these assumptions are indefensible. Some are undoubtedly more defensible than others. Instead, my aim has been to bring to light the background assumptions required to make this form of scientism truly threatening to philosophers and to emphasize their frequent strength and potential contentiousness, as well as the considerable work that would be required to adequately defend them. It should now be clear that strong epistemological scientism is not inherently threatening to philosophy or to philosophers; it is threatening only if we endorse some combination of the assumptions I have laid out.

3.2 Strong Methodological Scientism

I now turn to strong methodological scientism. Since I'm concerned with the potential threat of scientisms to philosophers, I'll address the formulation that concerns them directly: philosophers should only use the methods of science.

Note that this thesis requires *Methodological Distinctness* ((5) above), which posited distinctively scientific methods. Now, there's an immediate question as to the coherence of this thesis, because one might think that philosophers would cease to be philosophers if they exclusively used the methods of science.[11] If so, then the thesis seems to prescribe a contradictory state of affairs: philosophers doing something in virtue of which they are not philosophers (or doing *nothing* in virtue of which they *are* philosophers).[12,13] The problem stems from the thought that doing philosophy is essential to being a philosopher. In the interests of examining a prima facie coherent view, let's not assume as much.[14]

Why might one think that philosophers should use only the methods of science? One reason might be the belief that only scientific methods can produce desired epistemic goods. In other words, strong methodological scientism may naturally rest on strong epistemological scientism. Philosophers should use only the methods of science *because* that's the only way to get knowledge or whichever epistemic goods science has sole access to according to one's preferred version of strong epistemological scientism. If strong methodological scientism comes as a package deal with strong epistemological scientism, then the question of whether it is plausibly true and plausibly threatening to philosophers should be treated as the question of whether the package deal is plausibly true and plausibly threatening to them. Whether we think so depends on whether we accept some combination of assumptions (1)–(10) above, plus whichever additional ones the methodological thesis invokes.

Let's consider those additional assumptions. If science has a monopoly on certain epistemic goods (strong epistemological scientism), and those goods are the only genuine epistemic goods (as per (3) *Epistemic Value Monism*) or the only goods worth pursuing in academic contexts (as per (4) *Professional Value (Restrictive)*), and if distinctively scientific methods are required for the production of those goods (as per (6) *Scientific Indispensability*), then in order for it to follow that philosophers should use only scientific methods (strong methodological scientism) one must additionally assume:

(11) *Restricted Value Promotion*: Philosophers should promote only the epistemic goods on which science has a monopoly.

To synthesize the operative assumptions, the idea is that philosophers should exclusively use the methods of science *because* that's the only way to promote the epistemic goods on which science has a monopoly, which philosophers should exclusively promote *because* those are the only goods worth their time.

Sometimes methodological scientisms are formulated in terms of "empirical methods" or "the empirical methods of science." For instance, Mizrahi writes that scientism advocates "the use of empirical methods of observation, experimentation, and the like" (2019, 1). An empirical formulation of strong methodological scientism states that philosophers should exclusively use the *empirical* methods of science. Proponents of this empirical formulation may require additional assumptions, depending on how they view the relation between empirical and scientific methods. As we saw above, not all methods employed by science are empirical. If so, the thesis that philosophers should exclusively implement empirical methods is more specific than the thesis that they should exclusively implement scientific ones. If defenders of the empirical formulation acknowledge that the category of scientific methods exceeds the category of empirical methods, then their thesis requires the following assumption:

(12) *Unique Credit*: Only the empirical methods of science are implicated in the success of science vis-à-vis its proprietary epistemic goods.

We would need an assumption of this kind to justify using only the empirical methods of science. Note that *Unique Credit* is ambiguous between the claim that philosophers should exclusively use *all* of the empirical methods of science and the claim that they should exclusively use *some* of them. If the claim is that all of its empirical methods should be used, then it must be assumed that:

(13) *Total Credit*: All of the empirical methods of science are implicated in the success of science vis-à-vis its proprietary epistemic goods.

If the claim is that only some of the empirical methods should be used, then it must be assumed that:

(14) *Partial Credit*: Only some of the empirical methods of science are implicated in the success of science vis-à-vis its proprietary epistemic goods, and we know what they are.

The clause that *we know what they are* is needed because, on this interpretation, strong methodological scientism prescribes the exclusive use of those specific methods.

Proponents of the empirical formulation might avoid the need for these additional assumptions by denying that the category of scientific method exceeds the category of empirical methods. For instance, they might define "scientific method" as essentially empirical, such that any nonempirical methods scientists happen to use are by definition not genuinely scientific. However, they would have to provide a non-ad-hoc, non-question-begging rationale for defining scientific method so narrowly.

Does strong methodological scientism of the regular or of the empirical variety pose any special threat to philosophers, independently of its association with strong epistemological scientism? Let's again consider the statuses of philosophers, the value of their teaching and research, and their livelihoods. Strong methodological scientism threatens those things only if:

(15) *Professional Value (Permissive)*: For philosophers to be considered serious academics, for their teaching and research to be valuable, or for them to have a proper place as faculty members in universities requires it to be methodologically permissible for them to sometimes use non-scientific/non-empirical methods.[15]

Such non-scientific/non-empirical methods could include distinctively philosophical methods. On this assumption, for philosophers and their work to be respected and valued, it wouldn't be enough for them to implement scientific/empirical methods to address their own unique questions, even if they did so especially inventively or expertly, with interesting or fruitful results. Rather, the respectability and value of philosophers and their work hangs on philosophers' entitlement to implement other sorts of methods, perhaps including their own proprietary methods.

As for philosophers' preferred research methods, strong methodological scientism threatens to undermine them only for those philosophers described by the following assumption:

(16) *Non-Naturalism*: Some philosophers' preferred research methods exclude scientific/empirical methods.

This assumption is relatively uncontroversial. For those it describes, strong methodological scientism does—if true—undermine their preferred research methods, since it says they ought to use other methods exclusively. The question is roughly what proportion of philosophers fall under this category, and how many of them would agree to clumsy circumscriptions of philosophical and scientific methods according to which, for instance, armchair speculation is inherently unscientific and empirical research inherently unphilosophical.

Finally, strong methodological scientism threatens the health of the discipline only if:

(17) *Disciplinary Health (Permissive)*: The health of the discipline of philosophy requires it to be methodologically permissible for philosophers to sometimes use non-scientific/non-empirical methods.

Why think that disciplinary health requires it to be acceptable for philosophers to use non-scientific/non-nempirical methods? One reason might be the belief that philosophical questions cannot always be addressed by scientific/ empirical methods or cannot always be addressed by them alone. In other words:

(18) *Scientific Insufficiency*: Scientific/empirical methods are insufficient for the purpose of addressing philosophical questions.

If so, and if philosophers were permitted to use only scientific/empirical methods, they would make limited headway on those questions.[16]

An alternate rationale might be that if philosophers were to adhere to a methodology that proscribed the use of philosophical methods, philosophy would cease to be philosophy. This is related to (but not strictly the same as) the earlier thought that doing philosophy is essential to being a philosopher. This related thought is as follows:

(19) *Essentiality*: Distinctively philosophical methods are essential to philosophy.

The idea is that what makes philosophy *philosophy* is its own distinctive approach to addressing its subject matter. According to *Essentiality*, if we adhered to methodological norms proscribing that approach, philosophy would be lost. Such a view isn't immediately implausible but would require spelling out and defending. In particular, it would require the elucidation of philosophy's distinctive methods, as well as a defense of their essentiality.

We have seen that since strong methodological scientism is likely motivated by strong epistemological scientism, the truth and plausibility of the one requires some of the same assumptions as of the truth and plausibility of the other. Moreover, the claim that philosophers should exclusively employ scientific methods requires the assumption that philosophers should exclusively promote those goods that are uniquely valuable and that science is uniquely equipped to promote. We saw that methodological scientisms can be formulated in terms of scientific methods or in terms of empirical methods. Those who prescribe the exclusive use of empirical methods while granting that some scientific methods are non-empirical must say whether *all and only* or just *some of* the empirical methods of science are implicated in the relevant scientific successes. We also saw that strong methodological scientism threatens the status and value of philosophers and their work, as well as their jobs, only on the assumption that those things depend on the methodological permissibility of using non-scientific/non-empirical methods. Moreover, strong methodological scientism threatens only the preferred research methods of those who accept a clean circumscription of distinctively philosophical methods and distinctively scientific/empirical methods and who prefer the former. Finally, it threatens the health of the discipline only on the assumption that philosophy's disciplinary health requires it to be acceptable for philosophers to sometimes use non-scientific/non-empirical methods. Again, some of these assumptions are more prima facie plausible than others. The point is that some combination of them must be held fixed if strong methodological scientism is to be plausible and threatening to philosophers and that establishing them requires substantial work.

3.3 Strong Disciplinary Scientism

Finally, we have strong disciplinary scientism, according to which science will/should subsume/replace all other forms of inquiry. Let's start with the prospect of replacement, since "replace" has clearer connotations than "subsume." The prospect of science replacing philosophy makes sense only if the two are different things in the first place (as per (1) *Demarcation*). Once we assume so, the prospect of science *replacing* philosophy is threatening to philosophers—to their status, to the value of their teaching and research, or to their jobs—assuming the following:

(20) *Tribulations*: The replacement of philosophy by science entails the diminishment of philosophers' individual prestige, the cessation of their teaching and research, or the loss of their livelihoods.

The prospect of philosophy's "replacement" by science does seem to suggest the replacement of *philosophers* by *scientists*, an accompanying loss of prestige, and the cessation of their teaching and research (at least in institutional settings). Moreover, their teaching and research couldn't very well be valuable if it ceased to occur. So strong disciplinary scientism spelled out in terms of replacement does seem straightforwardly threatening to philosophers, without especially strong assumptions.

As for philosophers' preferred research methods, some of the assumptions I have already discussed are required to make strong disciplinary scientism threatening in that regard: there are distinctively philosophical methods ((5) *Methodological Distinctness*), some philosophers prefer methods such as those to scientific ones ((16) *Non-Naturalism*), and the replacement of philosophy by science may involve the cessation of philosophical research (*Tribulations*). The replacement form of strong disciplinary scientism threatens philosophers' preferred research methods only on those assumptions.

Lastly, replacement formulations of strong disciplinary scientism do clearly threaten disciplinary health. That is because the replacement of the discipline of philosophy entails its elimination. The discipline cannot very well be healthy if it ceases to be! So, as a thesis about replacement, strong disciplinary scientism either normatively or predictively opposes the very possibility of disciplinary health.

All in all, replacement formulations of strong disciplinary scientism do threaten philosophers without much need for strong or contentious assumptions. However, we might wonder how many avowedly scientistic thinkers actually endorse such an extreme form of scientism. Again, empirical research should support our sociological narratives, but I find it unlikely that many—otherwise sensible—people think philosophy should or will be consigned to the flames. So while this kind of strong disciplinary scientism poses a theoretical threat to philosophers in virtue of its content and some relatively reasonable assumptions, I don't believe it poses an immediate practical threat to them in virtue of being broadly believed (though empirical evidence could prove me wrong).[17]

The *subsumption* of philosophy by science, on the other hand, does not pose so obvious an existential threat. Whether the prospect of subsumption should truly threaten philosophers depends on what subsumption entails. The term is meant to suggest the extension of the boundaries of science over philosophy, such that philosophy becomes part of science. This could entail changes to the organizational structure of universities:

(21) *Departmental Subsumption*: The subsumption of philosophy under science entails that philosophy departments become part of faculties of science.

This would have practical consequences for course listings, program requirements, and so forth, but in being merely organizational, it would have no implications for our deeper questions about demarcation, epistemology, and methodology. Likewise, unless further conditions are specified (pertaining, for instance, to diminished budgets), it has few direct consequences for the professional standing of philosophers, the preservation of their methods, or the health of the discipline. In and of themselves, such organizational changes would harm philosophers professionally only if the following were true:

(22) *Professional Value (Departmental Autonomy)*: For philosophers to be considered serious academics, for their teaching and research to be valuable, or for them to have a proper place as faculty members in universities requires that departments of philosophy not belong to faculties of science.

It isn't clear why this would be so. Moreover, since the merely organizational characterization of subsumption has no methodological implications, it does not threaten to undermine philosophers' preferred research methods. It harms disciplinary health only on the assumption that:

(23) *Disciplinary Health (Departmental Autonomy)*: The health of the discipline of philosophy requires that departments of philosophy not belong to faculties of science.

Again, it's not clear why this would be so. So on a merely organizational conception of disciplinary subsumption, strong disciplinary scientism is not obviously threatening.

Alternate conceptions of subsumption could entail more substantive transformations to the way we think about philosophy and to philosophy itself. For example:

(24) *Substantial Subsumption*: The subsumption of philosophy under science entails that philosophical questions are properly scientific questions, which can/should be addressed using only scientific methods.

According to the first clause of this assumption, what some philosophers erstwhile thought to be at least partly an independent subject matter in fact belongs entirely to the domain of scientific interest and investigation. Such a view threatens philosophers professionally only if:

(25) *Professional Value (Substantial Autonomy)*: For philosophers to be considered serious academics, for their teaching and research to be valuable, or for them to have a proper place as faculty members in universities requires that philosophy have a subject matter at least partly independent of science.

This assumption would need defending. As for their preferred research methods, the second clause of *Substantial Subsumption*, which I'll address momentarily, speaks directly to method. Regarding disciplinary health, strong disciplinary scientism, paired with *Substantial Subsumption*, threatens the disciplinary health of philosophy only on the further assumption that:

(26) *Disciplinary Health (Substantial Autonomy)*: The health of the discipline of philosophy requires that it have a subject matter at least partly independent of science.

This assumption would likewise need defending. So where strong disciplinary scientism prescribes or predicts the subsumption of philosophy by science, and where subsumption is taken to entail that the subject matter of philosophy falls entirely under the purview of science, some meaty assumptions are required to make the view threatening to philosophers.

Different readings of *Substantial Subsumption* are continuous with different forms of strong methodological or epistemological scientism. For instance, the idea that only scientific methods *can* address putatively philosophical questions seems to follow from the thesis that science is the only source of evidence about certain sorts of question (one form of strong epistemological scientism), together with the assumptions that attribute science's epistemic successes exclusively to its proprietary methods ((5) *Methodological Distinctness* and (6) *Scientific Indispensability*). The idea that only scientific methods *should* be used to address our questions is equivalent to strong methodological scientism. If the subsumption formulation of strong disciplinary scientism invokes strong epistemological or methodological scientism, then the assumptions required to make those views plausible and threatening must also be invoked here.

In sum, as a thesis about the replacement of philosophy by science, strong disciplinary scientism threatens philosophers' professional standing and the health of the discipline of philosophy holding fixed relatively modest assumptions. Nevertheless, we may wonder just how broadly accepted such extreme forms of scientism are. As a thesis about the subsumption of philosophy by science, strong disciplinary scientism is less obviously threatening. If subsumption is understood to be merely organizational, then strong disciplinary scientism has rather dull teeth. If it is understood to be more substantively transformational, then strong disciplinary scientism threatens philosophers only assuming the importance of a fully independent subject matter, and only if we import some of the assumptions required to make strong epistemological or methodological scientism plausibly true and threatening.

I wish to pause at this juncture to consider the significance of what has been shown. We have seen that, for the most part, strong epistemological, methodological, and disciplinary scientisms must be supplemented by strong and often contentious assumptions in order to be truly threatening. Just one interpretation—the replacement interpretation of strong disciplinary scientism—was plausibly threatening under modest assumptions. This is striking! As I said at the beginning of the section, I focused first on the strongest forms of scientism, because if any forms are straightforwardly and acutely threatening, one would think it would be them! However, I have shown that this is largely not the case. For the most part, not even the comparatively extreme forms of scientism are terribly menacing in and of their own accord. I believe this is an important upshot. It would be far less surprising to find that more moderate forms of scientism aren't all that intrinsically menacing; after all, they're *moderate*! However, for the sake of completeness, it's still worth considering them—a task I turn to now.

4. WEAK SCIENTISM

I will take Mizrahi's *Weak Scientism* as a representative example of a more moderate form of scientism. According to Weak Scientism, "Of all the knowledge we have, scientific knowledge is the best knowledge" (2017, 354). Scientific knowledge is best, he claims, in terms of research output and impact, as well as with respect to its explanatory, instrumental, and predictive successes (2017, 356–357). By the lights of the characterizations of scientism I outlined in section 2, this "weak" scientism would count as a moderate form of epistemological scientism. However, I'll follow Mizrahi and others by calling it "Weak Scientism." Mizrahi intends Weak Scientism not to be inherently vicious (2017, 352); what sorts of assumptions would make it threatening to philosophers? As a weaker form of scientism, it shouldn't be surprising that making it plausibly true and plausibly threatening to philosophers will require assumptions that are, in some cases, even stronger than the ones we've seen so far.

Weak Scientism may be thought to have negative implications for the epistemic status of philosophy. For instance, if scientific knowledge is the best form of knowledge, then we might think philosophical knowledge is a lesser form. This follows only if we assume:

(27) *Distinctive Knowledge*: We can distinguish distinctively "scientific" from distinctively "philosophical" knowledge.

We might make the distinction by claiming that scientific knowledge is found in the journal articles, textbooks, manuscripts, volumes, and other venues of research dissemination that are classified as scientific, while philosophical knowledge is found in the ones classified as philosophical. This seems to be Mizrahi's preferred way of making the distinction when he considers things like impact factors. But it should go without saying that this way of making the distinction is conventional, historically contingent, and not necessarily philosophically well-founded.

Alternatively, we could say that scientific knowledge results from distinctively scientific methods, while philosophical knowledge results from distinctively philosophical methods. To do so, we would have to import (5) *Methodological Distinctness*, which distinguished the two forms of method. In that case, defending *Distinctive Knowledge* would require defending *Methodological Distinctness*.

Supposing we can cleanly distinguish scientific from philosophical knowledge, in order to conclude that philosophical knowledge is a lesser form of knowledge, we would also have to assume:

(28) *Exclusive Best-Making*: Scientific knowledge has features that philosophical knowledge lacks, in virtue of which it is best.

Mizrahi defends this sort of assumption when he points to the comparative differences between research output and impact in the sciences versus the arts and humanities (2017, 356–357), as well as differences in explanatory, instrumental, and predictive success between science and philosophy (2017, 358–362). The point is that defenders of Weak Scientism need to spell out and defend the appropriateness of certain measures by which we can judge knowledge better or worse and to show that science outperforms philosophy relative to those measures.

For Weak Scientism to undermine philosophers' prestige, the value of their teaching and research, or their livelihoods, we would have to assume the following:

(29) *Professional Value (No Inferiority)*: For philosophers to be considered serious academics, for their teaching and research to be valuable, or for them to have a proper place as faculty members in universities requires that philosophical knowledge not be a lesser form of knowledge.

But this isn't reasonable. Philosophers might be considered serious academics, their teaching and research might be valuable, and they might have a proper place as faculty members in universities even if philosophical knowledge were a lesser form of knowledge. That's because philosophical knowledge might still be valuable—even exceptionally valuable—even if it weren't the best form of knowledge. Moreover, philosophers might promote goods other than knowledge, both epistemic and non-epistemic. This shows that in order to truly undermine the status and value of philosophers and their work, something like (3) *Epistemic Value Monism* and (4) *Professional Value (Restrictive)*—which together ruled out the possibility of philosophy achieving anything worthwhile—would be needed. More specifically, we would have to assume:

(30) *Epistemic Value Monism+*: Scientific knowledge is the only genuine epistemic good.
(31) *Professional Value (Restrictive)+*: For philosophers to be considered serious academics, for their teaching and research to be valuable, or for them to have a proper place as faculty members in universities requires that they promote scientific knowledge.

These assumptions entail that promoting putative goods other than scientific knowledge is insufficient for philosophers and their work to be respected and valuable. They imply that in academic contexts, scientific knowledge is the only good *simpliciter* worth recognizing, promoting, and preserving. This is an exceptionally narrow view of what makes our academic activities valuable or worthwhile.

As for preferred research methods, it might be thought that the claim "scientific knowledge is best" negates the value of philosophical method. However, it does so only if all we care about is producing scientific knowledge ((3) *Epistemic Value Monism*), if there are distinctively scientific and distinctively philosophical methods ((5) *Methodological Distinctness*), and if only the scientific methods can get you scientific knowledge ((6) *Scientific Indispensability*). For it to follow that we should stop using philosophical methods altogether, we'd have to add:

(32) *Restricted Permissibility*: A method should be used only if it conduces to scientific knowledge.

The reason for thinking this might be that we want our method to promote valued goods, and scientific knowledge is the only valuable epistemic good or the only good *simpliciter* worth recognizing, preserving, or promoting in academic contexts (i.e., *Epistemic Value Monism+* and *Professional Value (Restrictive)+*). Overall, a substantial package of assumptions is required for Weak Scientism to threaten the value and continued advisability of using philosophical methods.

Finally, regarding the health of the discipline of philosophy, Weak Scientism threatens it only on the following assumption:

(33) *Disciplinary Health (No Supremacy)*: The health of the discipline of philosophy requires that scientific knowledge not be the best form of knowledge.

A natural corollary might add that philosophy should be the best form. Yet it is not clear why this would be so, when in principle Weak Scientism allows that philosophical knowledge might still be a good—even exceptionally good—form of knowledge. Even if philosophical knowledge were an especially poor form of knowledge according to the metrics we use to gauge such things (as is apparently the case in Mizrahi 2017), it wouldn't obviously condemn the health of the discipline. That's because there would still be a question as to why we should privilege those particular metrics and not others. Moreover, it would still be possible for philosophy to promote other epistemic and non-epistemic goods that our assessment of disciplinary health should consider.

To sum up, like the strong forms of scientism, Mizrahi's more moderate "Weak Scientism" poses an unintended threat to philosophers only holding fixed a number of meaty assumptions. Those assumptions must circumscribe philosophical and scientific knowledge and distinguish them in respects relevant to the bestness of the knowledge. Moreover, they must declare the production of scientific knowledge to be the only valuable academic end and the sole determinant of the worth of philosophers, their work, their methods, as well as the health of their discipline. This is certainly an extreme view, and if it has any defenders, they have their work cut out for them.

5. CONCLUSION

In this chapter, I set out to consider the sorts of assumptions required to make various scientisms threatening to philosophers, in the sense of potentially harming, disrupting, or undermining their prestige, the value of their teaching and research, their jobs, their preferred research methods, and the health of their discipline. In particular, I examined the assumptions required to make strong epistemological, methodological, and disciplinary scientism, as well as Mizrahi's Weak Scientism threatening in those respects. I found that most of the scientisms considered are neither straightforwardly nor inherently threatening to philosophers. Rather, most of them—including, tellingly, almost all of the strong scientisms—require the supplementation of substantive assumptions in order to be plausibly true and plausibly threatening to philosophers. The replacement form of strong disciplinary scientism was atypical in requiring the addition of relatively modest and uncontentious assumptions. In general, the assumptions surveyed heavy-handedly circumscribe philosophy and science, their epistemic credentials and achievements, their methods, and their subject matters. They also severely restrict the epistemic and non-epistemic goods considered valuable, worth promoting, and relevant to disciplinary health. While I have scarcely begun to fully address each assumption or its merit, I have revealed the numerous and substantive assumptions required for the discussed scientisms to be genuine specters. My hope for the future of the dialectic is that, in addition to consistently and carefully disambiguating "scientism," we exhibit greater awareness of and attention to the epistemological, methodological, and value-theoretic assumptions on which rest our attitudes toward particular forms of scientism.

ACKNOWLEDGMENTS

This research was supported by the Fundação para a ciência e a tecnologia, award #PTDC/FER-HFC/30665/2017. I thank Moti Mizrahi and the other contributors to the *Social Epistemology Review and Reply Collective* dialogue on sciences, which was the starting point for this chapter. For helpful feedback on a draft of this chapter, I thank David Yates and an anonymous reviewer.

NOTES

1. By this, I mean that we can stipulate meanings for terms of art. There are of course reasons why certain choices might be impractical or strange. For instance, it would be impractical to define "scientism" in a way that had nothing to do with how anyone else uses it. Moreover, it would be strange to define it in a way that had nothing to do with science. But even if it would be impractical or strange for us to do these things, we still could.

2. Though I acknowledge that this *ponens* could well be *tollens*-ed.

3. Similarly, Mizrahi argues that our definition shouldn't be *persuasive* in the sense of being inherently disapproving (2017, 352).

4. In characterizing methodological scientism prescriptively, I have followed Peels, according to whom "on methodological scientism, all or some academic disciplines different from the natural sciences should adopt the methods of the natural sciences in order to solve the problems of those fields" (2018, 49).

5. On the relation of scientism to physicalism, see Ney (2018); on the relation of scientism to philosophical naturalism, see Stenmark (2018); on the relation of scientism to realism, see Nickles (2017). On the confusion of scientism with interdisciplinarity, see Bishop (2019).

6. The variety of substantially different conceptions leading to confusion and cross-purposes suggests that "scientism" is *conceptually fragmented* in a manner described by Taylor and Vickers (2017). They argue for the elimination of fragmented concepts; however, I tend to think "scientism" can be used innocuously so long as we remain diligently conscientious.

7. While I formulate the assumption in general terms, knowledge will be my running example of science's allegedly proprietary epistemic good. However, we could equally well substitute another epistemic good, or more than one, in its place.

8. While this assumption is similar to *Demarcation*, which declared philosophy and science to be distinct, it is more specific. It is consistent with *Demarcation* that philosophy and science are distinct in virtue of something other than their methods (such as their subject matters).

9. Of course, it is disputed whether the less empirical pockets of science (such as string theory) count as bona fide science (see Castelvecchi 2015). Moreover, the alleged apriority of mathematics and logic is a matter of long-standing controversy (see, for instance, Kitcher 1983). So I grant that this is a quick and contentious way of making the point.

10. A defender of this position might have to appeal to something like the distinction between contexts of discovery and of justification in order to handle cases where speculation demonstrably led to discovery and advancement in science (such as well-known Einsteinian thought experiments).

11. Note that it's the strength of the thesis that generates this worry. Weaker formulations don't face the same problem, since it's completely coherent to imagine philosophers *sometimes* using nonphilosophical methods.

12. Alternatively, the thesis might be interpreted as the claim that philosophers should stop being philosophers and start being scientists, which improves on the previous interpretation with respect to coherence but remains remarkably strong.

13. I thank David Yates for raising this problem to me.

14. However, we will consider a related assumption (*Essentiality*) below, which, instead of claiming directly that method makes philosophers, will claim that method makes philosophy.

15. As above, the slash indicates possible alternate formulations—I'm not using it to equate "scientific" and "empirical," which I have already said is an ill-advised conflation.

16. This raises the vexed issue of whether and to what extent philosophers make progress on philosophical questions, anyway, as well as the question of how to measure such progress. I'll bracket those issues here.

17. If so, one might wonder why I have bothered to consider the replacement interpretation at all. As I mentioned at the start of section 3, I believe it is instructive to consider how *truly* threatening the most prima facie threatening forms of scientism are.

REFERENCES

Bishop, R. (2019) 'Scientism or Interdisciplinarity?', *Social Epistemology Review and Reply Collective*, 8(12), pp. 46–49.

Bourget, D. and Chalmers, D. (2014) 'What Do Philosophers Believe?', *Philosophical Studies*, 170(3), pp. 465–600.

Bryant, A. (2020) 'Keep the Chickens Cooped: The Epistemic Inadequacy of Free Range Metaphysics', *Synthese*, 197, pp. 1867–1887.

Buckwater, W. and Turri, J. (2018) 'Moderate Scientism in Philosophy', in de Ridder, J., Peels, R., and van Woudenberg, R. (eds), *Moderate Scientism in Philosophy*. Oxford: Oxford University Press.

Castelvecchi, D. (2015) 'Is String Theory Science?', *Scientific American*.

Chakravartty, A. (2013) 'On the Prospects of Naturalized Metaphysics', in Ross, D., Ladyman, J., and Kincaid, H. (eds), *Scientific Metaphysics*. Oxford: Oxford UP, pp. 27–50.

de Ridder, J. (2019) 'Against Empirical-ish Philosophy', *Social Epistemology Review and Reply Collective*, 8(12), pp. 8–12.

Dupré, J. (1993) *The Disorder of Things: Metaphysical Foundations of the Disunity of Science*. New York: Oxford University Press.

Field, H. (2005) 'Recent Debates about the A Priori', in Szabo Gendler, T. and Hawthorne, J. (eds), *Oxford Studies in Epistemology, Volume 1*. Oxford: Oxford University Press, pp. 69–88.

Haack, S. (2017) 'The Real Question: Can Philosophy be Saved?', *Free Inquiry*, 37(6), pp. 40–43.

———. (2012) 'Six Signs of Scientism', *Logos and Episteme*, 3(1), pp. 75–95.

———. (2003) *Defending Science - within Reason: Between Scientism and Cynicism*. Amherst: Prometheus Books.

Hietanen, J., Turunen, P., Hirvonen, I., Karisto, J., Pättiniemi, I., Saarinen, H. (2020) 'How Not to Criticism Scientism', *Metaphilosophy*, 51(4), pp. 522–547.

Kidd, I. J. (2019) 'Scientism and the "Soul of Philosophy"', *Social Epistemology Review and Reply Collective*, 8(11), pp. 52–54.

———. (2018) 'Is Scientism Epistemically Vicious?', in de Ridder, J., Peels, R., and van Woudenberg, R. (eds), *Scientism: Prospects and Problems*. Oxford: Oxford University Press, pp. 222–249.

Kitcher, P. (1983) *The Nature of Mathematical Knowledge*. Oxford: Oxford University Press.

Ladyman, J. (2018) 'Scientism with a Humane Face', in de Ridder, J., Peels, R., and van Woudenberg, R. (eds), *Scientism: Prospects and Problems*. Oxford: Oxford University Press, pp. 106–126.

Ladyman, J. and Don Ross, with David Spurrett and John Collier. (2007) *Every Thing Must Go: Metaphysics Naturalized*. Oxford: Oxford UP.

Mizrahi, M. (2019) 'The Scientism Debate: A Battle for the Soul of Philosophy?', *Social Epistemology Review and Reply Collective*, 8(9), pp. 1–13.

———. (2017) 'What's So Bad about Scientism?', *Social Epistemology*, 31(4), pp. 351–367.

Ney, A. (2018) 'Physicalism, Not Scientism', in de Ridder, J., Peels, R., and van Woudenberg, R. (eds), *Scientism: Prospects and Problems*. Oxford: Oxford University Press, pp. 258–279.

Nickles, T. (2017) 'Strong Realism as Scientism: Are We at the End of History?', in Boudry, M. and Pigliucci, M. (eds), *Science Unlimited? The Challenges of Scientism*. Chicago: University of Chicago Press, pp. 145–164.

Peels, R. (2018) 'A Conceptual Map of Scientism', in de Ridder, J., Peels, R., and van Woudenberg, R. (eds), *Scientism: Prospects and Problems*. Oxford: OUP, pp. 28–56.

Pigliucci, M. (2018) 'The Problem with Scientism', *Blog of the APA*. https://blog .apaonline.org/2018/01/25/the-problem-with-scientism/

———. (2010) *Nonsense on Stilts: How to Tell Science from Bunk*. Chicago: University of Chicago Press.

Rosenberg, A. (2020) 'Scientism Versus the Theory of Mind', *Social Epistemology Review and Reply Collective*, 9(1), pp. 48–57.

———. (2011) *The Atheist's Guide to Reality: Enjoying Life without Illusions*. New York: W.W. Norton.

Sorell, T. (1991) *Scientism: Philosophy and the Infatuation with Science*. London: Routledge.

Stenmark, M. (2018) 'Scientism and its Rivals', in de Ridder, J., Peels, R., and van Woudenberg, R. (eds), *Scientism: Prospects and Problems*. Oxford: Oxford University Press, pp. 57–82.

Taylor, H. and Vickers, P. (2017) 'Conceptual Fragmentation and the Rise of Eliminativism', *European Journal for Philosophy of Science*, 7, pp. 17–40.

Wilson, C. (2019) 'Regarding Scientism and the Soul of Philosophy', *Social Epistemology Review and Reply Collective*, 8(11), pp. 55–58.

Chapter 4

Conceptions of Philosophy and the Challenges of Scientism

Ian James Kidd

1. TWO HYPOTHESES

The recent surge of interest in scientism is a good opportunity to ask what is really at stake in debates about the nature and challenges of scientism in relation to philosophy. Moti Mizrahi has raised the question of the deeper metaphilosophical commitments and attitudes at work when people defend or resist scientistic attitudes and doctrines (Mizrahi 2019). One obvious feature of many of those debates is a peculiarly charged and often intemperate tone among their participants—accusations of arrogance and dogmatism between the participants to the debates and charges of obsolescence or hubris toward the attitudes and claims (Kidd 2018). Some intemperateness should be expected, of course, since many of us can, at times, be prone to overly hot-heated behavior. Still, deep chords are being struck, provoking strong reactions of a sort usually reserved for contentious moral and political topics.

Mizrahi is therefore right to invite us to pause and take stock of the state and the aims of philosophical debate. Stocktaking anyway makes sense, given all the new material added to the philosophical study of scientism lately, including two edited collections (de Ridder, Peels, and van Woudenberg 2018; Williams and Robinson 2016). There are also new studies of the anti-scientism of historical figures, notably Ludwig Wittgenstein (Beale and Kidd 2017). Moreover, periodic stocktaking helps us keep track of developing debates. A common feature of philosophizing is that definitions of problems change, something not always noticed by participants. Entrants to a debate often have different assumptions and motivations and only realize this once they are already embroiled in the back and forth.

A more specific reason to ask about the deep drivers of scientism debates is that they include attitudes and convictions of a metaphilosophical character that are both charged and contestable. Mizrahi nicely articulates this in terms of competing conceptions of the "soul of philosophy," by which he means, roughly, different ways of understanding the nature, values, and aims of philosophy. I think this is a good strategy. Many intellectual disagreements seem to be animated by deeper sensibilities and convictions that are at work just below the surface. Staying at the surface isn't always a bad thing—one can do a lot of work at the surface, and it is not always necessary to dive into the depths. Sometimes, though, the things we are trying to locate and explore are in the depths. If Mizrahi is right, making certain kinds of progress in the scientism debate will require grappling with deeper metaphilosophical issues.

A call to attend to the metaphilosophical issues driving debates about scientism was consistently made by the late Mary Midgley. An important critic of scientism, she is neglected in the current literature on scientism, to the loss of contemporary philosophical anti-scientism (Kidd 2015). Over some forty years, Midgley challenged many manifestations of scientism and urged us to attend to what she calls "myths"—not in the sense of false stories, but "imaginative visions" or background worldviews that shape ways of understanding the world and human life (see Midgley 1992, 2003). Sensitivity to "myths" requires critical self-reflectiveness and humility about the ways our attention and reflection can be affected by conditioning factors—psychological, intellectual, and cultural. Such factors should be investigated and then appraised, especially if they are—as many scientistic myths tend to be—"troublesome," "inconvenient," even "monstrous" (Midgley 2003: 132). In many cases, such appraisals will turn on motivating convictions about philosophy to issues concerning the "soul of philosophy."

Mizrahi offers two hypotheses about what scientism debates are really about. The first is that the scientism debate is a "battle for the future of philosophy as a discipline" (Mizrahi 2019: 1). By this he means something like organized forms of collective intellectual activity whose primary functions are teaching and research, manifested institutionally in departments, professional associations, journals, systems of training, and so on. Scientism, argues Mizrahi, threatens the integrity of philosophy as a discipline in various ways, not least by challenging or even denying its epistemic legitimacy. Unless kept in check, scientistic tendencies will gradually narrow the scope of legitimate work available to philosophers to zero. Certainly, champions of scientism urged this sort of dramatic aspiration to termination—from E.O. Wilson's famous call for the "biologisation of ethics" to more recent claims that neuroscience has resolved, once and for all, an array of issues in aesthetics and theology (see, for a critique, Tallis 2011).

Mizrahi's second hypothesis is that the scientism debate is, fundamentally, "a battle for the soul or essence of philosophy." Here we are in metaphilosophical territory. The soul or essence of philosophy is characterized in terms of what Mizrahi calls "traditional methods of philosophical inquiry . . . an *a priori* discipline" (Mizrahi 2019: 2). Scientism threatens the soul of philosophy by claiming that its methods and aims are better realized in or by the empirical sciences. A new, better form of disciplined inquiry is now fully realized—the sciences. Since retaining obsolete tools is wasteful, if not a danger to current workers, the better response is to cease philosophical activity.

An odd feature of Mizrahi's presentation of his two hypotheses is that both are offered as fundamental. Odd, since fundamentality is usually exclusive, yet we have two hypotheses. Presumably only one of them is actually fundamental and I think it is the metaphilosophical hypothesis about the "soul of philosophy." The reason is that the concern about the existence of the discipline of philosophy must presuppose an underlying set of claims about the nature and value of philosophy. Ultimately, philosophical teaching and research are valuable because they are the constitutive activities of an enterprise of the human spirit that *matters* in its own unique ways. Philosophy can matter in those ways because it has a distinctive "essence," one not shared with other intellectual endeavors, such as history or the natural sciences.

I suspect many philosophers feel the deep reason the topic of scientism matters is that it wrongly questions or impugns the integrity and significance of the discipline of philosophy. Often, debates proceed at two levels. At the surface, there are tussles about explanations of consciousness or the prospects for physicalism or the place for humanistic inquiries in a good life. At a deeper level, though, there are disputes about very fundamental issues about what it means to practice philosophy—for instance, the ambitions and values appropriate to an exercise of disciplined reflection and understanding which deserves to be called philosophical. Such metaphilosophical concerns may not always be at the forefront during debates about scientism. I think philosophizing about scientism is often at its best when engaging with more specific targets—maybe a dubious "neurotheological" claim, perhaps, or some "demolition" of free will running on some mangled form of physicalist reductionism. Sometimes, though, we should engage much broader metaphilosophical issues directly.

2. CRUDITY AND CHALLENGES

Calls to engage in metaphilosophical reflection are not always met with approval. Doubtless, there are many philosophers in the scientism debate who will want to resist them. Granted, much excellent philosophical work on scientism does not engage directly metaphilosophical issues and is none the worse for it. My claim is only that those issues are usually there in the background and that sometimes we do well to engage with them. To that end, consider two general points to bear in mind when getting metaphilosophical about scientism.

First of all, there are many different reasons—cultural, sociological, ideological—for contemporary attacks on the value and integrity of philosophy. Only some of these turn on anything as complex as entrenched dogmas about the authority of the sciences relative to philosophy and other humanistic disciplines. In many cases, modern foes of philosophy are motivated by far more quotidian considerations. Other attacks on philosophy come from those who only value academic subjects they regard as economically valuable, those ideologically hostile to culturally élite disciplines, or those who misconceive philosophy as an aimless discipline incapable of making any tangible difference to the world. Some of these foes sometimes appeal to scientism: a whole genre of books now exists of critical studies of academia and higher education, whose *bête noires* often include intruding forms of scientism. An ominously titled book, *The University in Ruins*, envisions a future academia in which the natural sciences are regarded as purveyors of "real knowledge and large toys," while reducing the humanities to "cultural manicure" (Readings 1996, 172).

It is obvious that the threats to philosophy are both numerous and deeply entangled, due to the contingencies of history and the dialectics of culture, politics, and history. I think we do best to remain alert to the various threatening "isms"—including scientism, philistine instrumentalism, profit-driven neoliberalism, and more besides. Mizrahi is right to target scientism, but it is only one of the threats to the integrity and future of philosophy. Scientism might not even be the most powerful in many of the countries in which philosophy and other humanistic disciplines are under threat. What's really needed, here, are careful histories of scientism, a rare example of which is Richard Olson's study of European scientism from Revolutionary France, Romantic Germany, and Victorian England (Olson 2008). Indeed, some champions of scientism regard themselves as engaged in a *defense* of the integrity of philosophy against what they see as its corruption at the hands of invidious developments—a point I return to shortly.

A second comment about engaging the metaphilosophical dimensions of debates about scientism is that they may sometimes afford the topic a depth of thinking it does not deserve. A fairly blunt response to scientism is that it is—quite simply—*wrong* and too obviously so to merit serious attention. Certainly, some forms of scientism are so crude that challenging them is too easy and some claims made by self-affirmed champions of scientism are simply false—for instance, Steven Hawking once castigated philosophers (all of them, apparently) for having not kept up to date with modern physics, a claim that doubtless came as a shock to many philosophers of physics (Kidd 2011).

I sympathize with those who urge us not to waste time on cruder forms of scientism. Unfortunately, the cruder forms attract attention and often find their way into the public sphere: recall the spats between some physicists and philosophers in several American and British newspapers a few years back, provoked by some critical reviews of a self-describedly scientistic book by the physicist Lawrence Krauss (Corneliussen 2012). Recall, too, that many of the various popular physicists who engage in science communication apparently find it difficult to resist the temptations of scientism. Hawking was only one of a gaggle of scientistic physicists with a high public profile.

A further problem with the advice not to engage with crude scientistic claims is that there are often trickier questions about what counts as the cruder and more sophisticated forms of scientism. Clearly, some forms of scientism are more sophisticated than others, by very general intellectual standards of argumentative ability, scholarly rigor, and coherence. Indeed, there are sophisticated forms of philosophical scientism developed by distinguished philosophers of science that deserve to be taken seriously, such as James Ladyman and Don Ross' self-declaredly polemical manifesto, *Every Thing Must Go*. They are good examples of avowedly scientistic philosophers acting, as they see it, in the best interests of the discipline of philosophy, especially analytic metaphysics, the main quarry. In their book, they defend a "radically naturalistic metaphysics . . . motivated exclusively by attempts to unify hypotheses and theories that are taken seriously by contemporary science" (Ladyman and Ross 2007: 1). As well as being philosophically serious and sophisticated, they are sincere in their concern to help contemporary theorists navigate safely the "widespread unscientific and even anti-scientific intellectual waters" in which we now swim (Ladyman and Ross 2007: 310). The characterization of their project makes clear the metaphilosophical dimensions of the topic of scientism. In the opening lines, Ladyman and Ross lay out their claim that "contemporary analytic metaphysics . . . fails to qualify as part of the enlightened pursuit of objective truth, and should be discontinued":

> We think it is impossible to argue for a point like this without provoking some anger [. . .] Let us therefore stress that we wrote this book not in a spirit of hostility towards philosophy or our fellow philosophers, but rather the opposite. We care a great deal about philosophy, and are therefore distressed when we see its reputation harmed by its engagement with projects and styles of reasoning we believe bring it into disrepute, especially among scientists. (Ladyman and Ross 2007: vii)

Here we find a clear statement of the metaphilosophical dimensions of scientism of a sort that cannot be dismissed as intellectually crude. It's also a scientistic doctrine that is clearly entangled with other threats to philosophy, not least given the talk of "discontinuing" whole research programs that fall outside a naturalistic orthodoxy.

The upshot is that there are sometimes sophisticated forms of scientism that should be taken seriously and engaged with critically by philosophers concerned enough to engage them. Sometimes their concerns will be very specific ones pertaining to particular doctrines or debates—in metaphysics or philosophy of mind, say. Sometimes, though, such specific concerns are bound up with much more general issues about the methodology and goals of philosophy. At this point, one is getting caught up in the deeper metaphilosophical currents and swept toward the sorts of fundamental issues which Mizrahi articulates in terms of the "soul" or "essence" of philosophy—something he is keen to protect.

3. THE "ESSENCE OF PHILOSOPHY"

Talk of the "essence" of philosophy invites instant problems. An obvious one is that there are many different styles of philosophizing and a similarly large array of topics and at least three major world traditions—the Western, Indian, and Chinese, each evincing enormous internal diversity. Some law of comparative philosophy ensures that counterexamples will abound to any reasoned efforts to postulate some common or universal feature, even one moderated by assurances about family-resemblance concepts. I think it is wiser to speak in terms of a very broad array of *visions* of philosophy—some complementary, others conflicting—which have been realized in various forms across different cultures and traditions. A list of some general examples shows the varieties: philosophy as "spiritual practice," as a means of release from the "wheel of suffering," as "underlabourer" for the sciences, as a cure for "mental cramps," as an engine of progressive social change, as "conceptual geography," as a diverting cognitive game played for its own sake—and many others of similar sorts (see Cooper 2009). Some of these, like that of philosophy as a spiritual practice, go well beyond what Rik Peels calls "academic scientism," which is the concern of Ladyman and Ross. We can usefully explain the difference by appealing to Cooper's grouping of conceptions of philosophy into two main types. First, those for which philosophy is "an essentially theoretical, speculative enterprise," oriented "necessarily and primarily towards Truth," whether about reality [or] the conceptual schemes we employ for describing reality. Second, those for which philosophy is "a practical, vital enterprise," morally and spiritually charged, oriented "towards the Good, towards Life as it should be" (Cooper 2009: 1–2). Ladyman and Ross operate with a specifically naturalist conception of philosophy oriented toward Truth, according to which our best source of knowledge about reality is the natural sciences. According to that naturalistic conception, certain of our vital, practical needs will be either unintelligible—soteriological ones, say—or better served through other sorts of activities. "Philosophy as a way of life" will simply seem too highfalutin', since the guidance one needs for the proper conduct of life need not come from anything as ambitious as philosophical ways of life of the sort exemplified by Buddhism or Epicureanism and other ancient schools.

It should be clear that scientism will only appear as a threat to philosophy on some of these conceptions. Granted, much will depend on how "scientism" is characterized. There are many options, some offered in important earlier work on scientism by Tom Sorell (1991) and Mikael Stenmark (2001). Mizrahi's account of Weak Scientism entails the commitment is to the epistemological thesis that "the scientific way of knowing is the best way of knowing," in the sense that, "of all the knowledge we have, scientific knowledge is the best knowledge" (Mizrahi 2017). Such Weak Scientism, he argues, is necessary to ensure that philosophy must honor the authoritative epistemic achievements of the sciences if it is to remain a serious discipline. Mizrahi's definition was criticized for being too narrow by Christopher Brown (2017), a worry compounded by Rik Peels' account of at least three other extant forms of epistemic scientism (Peels 2018: 34). Broader conceptions of scientism are also available, culminating in forms of what Stenmark calls *comprehensive scientism* which includes epistemological, metaphysical, axiological, and existential theses and so "contains probably all other forms of Scientism that we have identified" (Stenmark 2001: 15). If defined very narrowly, scientism may not arouse any particular worries—if, say, its constituent theses pertain only to very narrow domains or if the contents of those theses are too circumscribed to be contentious. Yet as the claims grow broader, so, too, does the potential for alarm.

A similar point can be made about the scope and content of whatever conceptions of philosophy are in play when we debate scientism. At the absolute broadest, there is Wilfred Sellar's famous line that the aim of philosophy is "to understand how things in the broadest possible sense of the term hang together in the broadest possible sense of the term" (Sellars 1962: 37). At the other end, there are much more defined conceptions of philosophy as the analysis of ordinary language or—crossing the Channel—an articulation of the conditions of experience under the "disenchanted" conditions of late modernity. Looking to other cultures, philosophy also appears as the effort to liberate human beings from the cycle of karma and rebirth or the reflective commitment to achieve a state of harmonious "consummate ease."

The sheer abundance of conceptions of philosophy could feed a worry that the task is too large to attempt, at least if we proceed piecemeal. Fortunately, we can make our life easier by introducing a few categories, the most important being conceptions of philosophy that are intrinsically scientistic—like that of Ladyman and Ross'—and, relatedly, those that are intrinsically anti-scientistic. A good candidate is existential phenomenology as practiced by Martin Heidegger. In *Being and Time*, he argues that cognition "presupposes [human] existence," the shared forms of experience and activity, characteristic of creatures like us, which are themselves grounded in the "primordial" forms of experience and embodied engagement with the world. Our "being-in-the-world" has a thoroughly "intentional" form: we can only experience things in terms of the *significance* they enjoy relative to our values, interests, and projects—our being-in-the-world (Heidegger 1962: 15–16). Consequently, the natural sciences, though important for certain purposes, have an epistemologically derivative status: they cannot reveal how the world is independent of our experience and engagement, since they can only get to work on objects and processes already revealed within those primordial forms of engaged experience. (For a clear statement of this general form of anti-scientism running through the existential, phenomenological tradition, see Ratcliffe 2013).

Into their respective later periods, Heidegger and Edmund Husserl extend their own epistemological criticisms of scientism into a much broader set of cultural and existential concerns. Heidegger's influential later essays, *The Question Concerning Technology*, argue that our estimations of the powers of the natural sciences are part and parcel of the accelerating entrenchment of a baleful "technological" stance on the world driving the destruction of the natural world, the erosion of a vital sense of mystery, and the alienation of human beings from themselves, other persons, and the world (Heidegger 1977: 27ff and 118ff; see Cooper 2005). Whatever one thinks to such expansive criticisms of scientism, it's clear that hostility to scientism was an abiding, evolving theme in Heidegger's writings and that combating it was among his central concerns—for what is really "messing up" modern thought and culture, he declared, is "the dominance and primacy of the *theoretical*," which is most fully expressed in the entrenchment of scientism (Heidegger 1987: 87). Here, then, we have a good example of an intrinsically anti-scientistic conception of philosophy, one whose dominant *mood* or *sensibility*—as well as its doctrines and theses—are characterized by an enduring hostility to scientism, albeit not to science itself (cf. Peels 2018: 29).

I think that the category of intrinsically anti-scientistic conceptions of philosophy can be useful for organizing our thinking about the metaphilosophical issues in debates about scientism. It suggests, of course, that some conceptions are only *contingently* anti-scientistic. Consider the criticisms of evolutionary psychology developed in the work of John Dupré: the targets are evolutionary psychology and rational choice theory in the forms they had taken during the 1990s, ensnared and corrupted by the "lure of the simplistic"

(Dupré 2001, 2002). Had those disciplines developed in other ways, there would be no need to develop doctrines of epistemic pluralism specifically intended to challenge forms of imperialist scientism—that being an explicit metaphilosophical concern driving Dupré.

I suggested that talk of the "essence" of philosophy may become more tractable if we speak instead of a rainbow variety of conceptions of philosophy, which could be intrinsically or contingently scientistic or anti-scientistic. Ladyman and Ross offer an intrinsically scientistic conception of philosophy as naturalized metaphysics in deference to the deliverances of the contemporary physical sciences that excludes other styles and projects of inquiry from "the great epistemic enterprise of modern civilization." By contrast, Heidegger and the existential phenomenologists offer an intrinsically anti-scientistic conceptions of philosophy whose aim is a careful description of the fundamental structures of sensibility constitutive of our ways of experiencing and engaging with the world—the main obstacle to which is the conceit that the sciences can provide direct accounts of the world independent of those ways of "being-in-the-world." Between those two complex examples, one finds a whole variety of other ways of conceiving of the nature and significance of science in relation to the aims and purposes of philosophy.

4. GOING FORWARD

I started by endorsing Mizrahi's suggestion that those interested in scientism should look at the underlying metaphilosophical dimensions—deeper convictions about the nature, aims, and significance of philosophy. Without prescinding from other ideological and cultural worries, claims and counter-claims about scientism do often turn on rivaling visions of philosophy. It is important that those visions are described and presented upfront, rather than left lurking in the background without proper critical scrutiny. One natural difficulty of this sort of work is the variety of forms of scientism and the similar variety of conceptions of philosophy. This can help to explain the peculiarly charged character of debates about scientism—what is at stake is the integrity of the intellectual and cultural enterprise which has a definitive role in the self-identity of many practicing philosophers. While some will be sanguine in accepting their role as ontological bookkeepers for the sciences, others regard such deflated roles as a willful abandonment of the core purposes of the philosophical enterprise—a willfulness that for some borders on gross recklessness. Some philosophers will doubtless roll their eyes at such apocalyptic claims and scoff at Edmund Husserl's ruminations on "barbarian hatred of spirit" and its connection to the "crisis of the European sciences" (Husserl 1970: 299ff). But a scoff is worth much less than a thought. A better response to unfamiliar and expansive ways of conceptualizing the relations of philosophy to science and human life is to sit down to do the work of understanding them—to try and draw out what Midgley called the "imaginative visions" that serve to make them intelligible and compelling.

The upshot is that many philosophical debates about scientism are often the surface-level expressions of much deeper conflicts of a metaphilosophical character. Not always, for sure, since a lot of anti-scientism is driven by epistemological and practical issues rooted in the special authority of scientific knowledge and institutions in human life. Dupré, Tallis, and other critics of scientism often engage perfectly tractable worries about abuses of science in economics, healthcare, the arts, and so on. In other cases, though, really engaging with the issues means engaging with the operative claims about the nature and aims of philosophy. It matters that while for some scientism is a fundamental threat to the integrity of philosophy, to others it appears as a necessary step to its enhancement or final consummation. Wesley Buckwalter and John Turri, for instance, argue that experimental philosophy has enhanced, among other disciplines, ethics, epistemology, and philosophy of action (Buckwalter and Turri 2018).

A useful way forward in debates about scientism is to attend directly to these deeper claims about the "essence" or "soul" of philosophy and thereby establish just what is really at stake. We can take as our guide Mary Midgley's own reflections on the "myths" and "visions" which organize and animate our ways of understanding science, philosophy, and human life (see, e.g., Midgley 1993, 2003). A key theme of her work is insistence on proper appreciation of the respective powers and limits of science, philosophy, and other disciplines. We should, she argues, strive to maximize our epistemic resources through judicious employment of the many disciplines, traditions, and ways of thinking at our disposal. Some questions are best left to the sciences, others to philosophy, while others—perhaps most—require the careful use of resources of many different sorts. What Midgley urges is a careful, discerning, and pragmatic attitude toward philosophy and the sciences—accepting Buckwalter and Turri's claim that the sciences, used properly, can enhance philosophy, while avoiding those more radical proposals to "discontinue" whole areas of philosophy.

I endorse Midgley's cautious, particularist approach to science and philosophy, which is exemplified in her own work. Perhaps her most famous book, *Beast and Man*, was closely informed by ethology and evolutionary biology, a pattern of engagement with the sciences that continued into such later works as *Animals and Why They Matter*. "We do not need to esteem science less," she writes, "What we need is to esteem it in the right way" (Midgley 1992: 37). The same is true of philosophy, of course, which is only equipped to deal with certain sorts of problems: there are risks in inflating the scope and power of philosophy just as much as in doing the same for the sciences. Midgley's insistence on balancing proper respect for the contributions of the science with similar respect for philosophy and other disciplines is made clear in a remark from her final book. Speaking of the sorts of sensible, judicious ways of thinking she admires, she says:

> Their aim is always to help us through the present difficulty. They do not compete with the sciences, which at present supply our most dominant visions of reality. Instead, philosophy tries to work out the ways of thinking that will best connect these various visions—including the scientific ones—with each other and with the rest of life. (Midgley 2018: 6)

Our goal should be a conception of philosophy which preserves its distinctive functions and significance, but which also does the same for the sciences—to respect the essence of both, not just to avoid epistemological tangles, but also as a means of doing them both justice.

ACKNOWLEDGMENTS

I am grateful to the editor for the invitation to contribute to this chapter and to an anonymous referee for very helpful comments.

REFERENCES

Beale, Jonathan and Ian James Kidd, eds. 2017. *Wittgenstein and Scientism*. London: Routledge.

Brown, Christopher M. 2017. 'Some Objections to Moti Mizrahi's 'What's So Bad about Scientism?'' *Social Epistemology Review and Reply Collective* 6.8: 42–54.

Buckwalter, Wesley and John Turri. 2018. 'Moderate Scientism in Philosophy.' In Jereon de Ridder, Rik Peels and René van Woudenberg (eds.), *Scientism: Prospects and Problems*. Oxford University Press, 280–299.

Cooper, David E. 2005. 'Heidegger on Nature.' *Environmental Values* 14: 339–351.

Cooper, David E. 2009. 'Visions of Philosophy.' *Royal Institute of Philosophy Supplement* 65: 1–13.

Corneliussen, Steven T. 2012. '"Physicists, Stop the Churlishness": An Intellectual Conflict with Philosophy Draws Media Attention.' *Physics Today*, 13 June.

de Ridder, Jeroen, Rik Peels, and René van Woudenberg, eds. 2018. *Scientism: Prospects and Problems*. Oxford: Oxford University Press, 222–249.

Dupré, John. 2001. *Human Nature and the Limits of Science*. Oxford: Oxford University Press.

Dupré, John. 2002. 'The Lure of the Simplistic.' *Philosophy of Science* 69: S284–S293.

Heidegger, Martin. 1962. *Being and Time*, trans. J. Macquarie and E. Robinson. Oxford: Blackwell.

Heidegger, Martin. 1977. *The Question Concerning Technology and Other Essays,* trans. William Lovitt. New York: Harper & Row.

Heidegger, Martin. 1987. *Gesamtausgabe,* vol. 56/57. Frankfurt a. M.: Klostermann.

Husserl, Edmund. 1970. *The Crisis of European Sciences and Transcendental Phenomenology,* trans. David Carr. Evanston, IL: Northwestern University Press.

Kidd, Ian James. 2011. 'Three Cheers for Science and Philosophy! Reflections on Hawking's *The Grand Design.' Think: Royal Institute of Philosophy* 10: 37–41.

Kidd, Ian James. 2015. 'Doing Science an Injustice: Midgley on Scientism.' In Ian James Kidd and Elizabeth McKinnell (eds.), *Science and the Self: Animals, Evolution, and Ethics: Essays in Honour of Mary Midgley.* London: Routledge, 151–167.

Kidd, Ian James. 2018. 'Is Scientism Epistemically Vicious?' In Jeroen de Ridder, Rik Peels, and René van Woudenberg (eds.), *Scientism: Prospects and Problems.* Oxford: Oxford University Press, 222–249.

Ladyman, James and Don Ross. 2007. *Every Thing Must Go: Metaphysics Naturalised.* Oxford: Oxford University Press.

Midgley, Mary. 1992. *Science as Salvation: A Modern Myth and its Meaning.* London: Routledge.

Midgley, Mary. 2003. *The Myths We Live By.* London: Routledge.

Midgley, Mary. 2018 *What is Philosophy For?* London: Bloomsbury.

Mizrahi, Moti. 2017. 'What's So Bad About Scientism?' *Social Epistemology* 31 (4): 351–367.

Mizrahi, Moti. 2019. 'The Scientism Debate: A Battle for the Soul of Philosophy?' *Social Epistemology Review and Reply Collective* 8 (9): 1–13.

Olson, Richard. 2008. *Science and Scientism in Nineteenth-Century Europe.* Urbana and Chicago, IL: University of Illinois Press.

Peels, Rik. 2018. 'A Conceptual Map of Scientism.' In Jeroen de Ridder, Rik Peels, and René van Woudenberg (eds.), *Scientism: Prospects and Problems.* Oxford: Oxford University Press, 28–56.

Ratcliffe, Matthew. 2013. 'Phenomenology, Naturalism, and the Sense of Reality.' *Royal Institute of Philosophy Supplement* 72: 67–88.

Readings, Bill. 1996. *The University in Ruins.* Cambridge, MA: Harvard University Press.

Sellars, Wilfred. 1962. 'Philosophy and the Scientific Image of Man.' In Robert Colodny (ed.), *Frontiers of Science and Philosophy.* Pittsburgh, PA: University of Pittsburgh Press, 35–78.

Sorell, Tom. 2013. *Scientism: Philosophy and the Infatuation with Science.* London: Routledge.

Stenmark, Mikael. 2001. *Scientism: Science, Ethics, and Religion.* Aldershot: Ashgate.

Tallis, Raymond. 2011. *Aping Mankind: Neuromania, Darwinitis, and the Misrepresentation of Humanity.* Durham: Acumen.

Williams, Richard N. and Daniel N. Robinson, eds. 2016. *Scientism: The New Orthodoxy.* London: Bloomsbury.

Chapter 5

How to Defend Scientism

Petri Turunen, Ilkka Pättiniemi, Ilmari Hirvonen,
Johan Hietanen, and Henrik Saarinen

Recently, the discussion concerning scientism seems to have managed to move on from rhetorical dismissals and blanket statements. Instead of focusing on whether scientism is laudable or lamentable by definition, now the debate considers what scientism entails and what different forms it can take. In addition, Moti Mizrahi (2019), a defender of scientism, has considered the role of scientism in philosophy, why it is opposed, and the proper nature of philosophical inquiry. Of particular importance is what has come to be called *Weak Scientism* (Mizrahi 2017; Hietanen et al. 2020). Weak Scientism states that science is not the *only* known source of justification, evidence, or some other epistemic good—like Strong Scientism suggests—but merely the *best* one.

This chapter will take a closer look at Mizrahi's views on Weak Scientism and some of their problems. Moreover, to amend these shortcomings, we shall offer a defense of Weak Scientism based on epistemic opportunism. We start by examining Mizrahi's claim that philosophers' opposition to scientism is founded on the worry that scientism poses "a threat to the soul or essence of philosophy as an *a priori* discipline" (Mizrahi 2019, 2). We find Mizrahi's methodology for testing this thesis wanting. In addition, we point out that there can be alternative hypotheses for the increased resistance to scientism. One such alternative is that the antipathy started as a reaction to the New Atheist movement, which insisted that science renders theism unjustified.

After this, we will consider two different varieties of Weak Scientism: narrow and broad. Here we argue that narrow versions of scientism—that exclude disciplines like history as proper sources of knowledge—are problematic and draw artificial and unfounded distinctions within science. Mizrahi belongs somewhere between the narrow and broad types, but he commits the same mistakes as proponents of the narrow variety. Next, we look at Mizrahi's

(2017) defense of Weak Scientism and demonstrate its inadequacies. Here, again, one can find significant problems in Mizrahi's methodology. As an alternative, we propose that Weak Scientism should be based on epistemic opportunism. Epistemic opportunism explains the success of science with scientists' willingness to adopt any methods that demonstrably work, and this is how science *should* be practiced. We also show how scientism can avoid charges of triviality. Finally, we note that the scientistic critique of philosophy often pertains to the evaluability of philosophical methods. Based on this, we consider some implications for the relationship between philosophy and science and for the soul of philosophy.

1. SCIENTISM AND THE SOUL OF PHILOSOPHY

Mizrahi has recently employed his "scientistic" methodology to a metalevel question about scientism, namely, is the controversy over scientism "a battle for the soul of philosophy?" (Mizrahi 2019). He makes two tentative observations: First, the battle is over the future of philosophy as an independent field of study (id., 1). Second, the dispute is over what philosophy *is* or *should be* (id., 2). To empirically test these claims, he makes the following two hypotheses:

> H1: Many philosophers find scientism threatening because they see it as a threat to the future of philosophy as a major in colleges and universities (id., 2).

> H2: Many philosophers find scientism threatening because they see it as a threat to the soul or essence of philosophy as an a priori discipline (id., 2).

To test H1, Mizrahi studied the relationship between the number of undergraduates majoring in philosophy or religious studies over time and the number of philosophy publications mentioning "scientism" during the same period. He also did the same for theology and religious vocations BAs and mentions of "scientism" in religious publications. He reports that his findings do not support H1 (id., 8–9). But the test is built on a controversial premise, namely that "the more concerned philosophers [. . .] are about scientism, the more [publications] they would write about" it and the threat it poses (id., 4). This is supported by three instances of scholars writing about things that concern them.

To test H2, Mizrahi compares the number of publications in philosophy containing the phrase "experimental philosophy" with those with the term "scientism." Here he considers his findings to support H2 enough to warrant further investigation.

Be that as it may, there is a worry about whether Mizrahi is using the best or even good methods to investigate his hypotheses, whether he has employed those methods well, and whether he has formulated his hypotheses in a way that could be subjected to testing. Here Jeroen de Ridder's (2019) reply to Mizrahi (2019) is quite helpful. As de Ridder (2019, 9) points out, the hypotheses are rather vaguely worded: they talk of "many philosophers," but how many are many? They also refer to beliefs and feelings, and while one might use bibliometric methods to study a lot of subjects, using them to suss out a relation between propositional attitudes and behavior like academic publishing seems strained at best (id., 8–9). Also, as de Ridder points out, Mizrahi does precious little to rule out alternative hypotheses that would account for his data (id., 9). And further, even *if* the previous problems were taken care of, there is still the problem of whether we should infer any causal relations from the correlations in Mizrahi's data. Thus, his hypothesis can hardly be used as both a question under study and a reason to infer causation (id., 10–11).

In addition to de Ritter's worries, there is also another one that we alluded to earlier. Do scholars, when concerned about something, really write about it (Mizrahi 2019, 3)? As Mizrahi's three references show, it seems clear enough that they *sometimes* do so, but this seems rather weak. It is equally clear that scholars write about things they are *interested* in, regardless of whether they are also concerned about them. Might one of the reasons behind the surge of writings about scientism be a resurgence of metaphilosophy as a subject?

And there are alternative hypotheses to H1 and H2, which would explain the increased interest in scientism. Let us consider one briefly. At about the same time as experimental philosophy started to become a serious contender for philosophical methodology (Knobe and Nichols' "An Experimental Philosophy Manifesto" was published in 2008), the movement known as *New Atheism* was also getting its start, with the publication of Sam Harris' *The End of Faith* in 2004 and Richard Dawkins' *The God Delusion* in 2006. Proponents of New Atheism declare it to be a new scientifically informed form of atheism (see, e.g., Stenger 2009). So, might it be that the recent interest in scientism came about from an opposition to the apparent scientism of the New Atheists?[1] At least Massimo Pigliucci (2013) has criticized the New Atheists for their scientism. It may well turn out that there is no single explanation to be had, as is often the case with complex phenomena, but this will be left as an open question. In any case, it is possible that the increased interest in scientism might not be predominantly connected to the "battle for the soul of philosophy" but, instead, to other worldview debates—such as theological ones.

Still, there is something to Mizrahi's idea that the debate might be over the soul of philosophy, at least if by "soul" one means "the methods" of philosophy. And on precisely that point, it seems that there is confusion about what the methods of *science* are. For instance, when science is concerned, the division between "the armchair" and "the laboratory" seems ill-informed. How so? Because much of science, the *theoretical* bits, can quite well be done from the armchair. There are, of course, empirical constraints to theoretical science, but that does not force the theoretician into the laboratory. So, whatever the methodological divide is, this cannot be it. But to get closer to that question, we should first take a look at what a weak form of scientism is—or could be.

2. TYPES OF WEAK SCIENTISM

Mizrahi defines Weak Scientism as follows: "[o]f all the knowledge we have, scientific knowledge is the *best* knowledge" (Mizrahi 2017, 354 emphases in the original). This definition contrasts Weak Scientism with Strong Scientism. However, in Hietanen et al. (2020), we added another dimension to the classification of scientism, namely the distinction between Broad and Narrow Scientism, leading to four possible versions of scientism. The divisions are made so that the relevant types of scientism are represented by the cross-sections of two categories: strong/weak and broad/narrow. Here, we will only consider the two kinds of scientism under the "weak" category.

In Hietanen et al. (2020), we argue that there are two main types of Weak Scientism: narrow and broad. The narrow type states that the *natural* sciences are the best sources of knowledge, justification, and the like. On the other hand, the broad variety claims that not only the natural sciences but sciences *in general* are the best sources for the mentioned epistemic goods or something akin to them. So, the proponent of Broad-Weak Scientism would understand "science" along the lines of the German term *Wissenschaft*. It is good to note that it is possible to make the narrow/broad distinction differently than we have done here. For instance, someone could, in theory, insist that, say, history is the only proper science. However, here we simply follow the prevalent practice.

It is somewhat difficult to state which category of scientism Mizrahi belongs to.[2] His definition of science predominantly emphasizes the natural sciences, but he also accepts the social sciences as sciences (Mizrahi 2017, 356–58). Still, despite this, Mizrahi is recalcitrant to recognize the humanities as proper sciences.

Some attention should be now given to the preference between these two types of Weak Scientism. We argue that scientism should not be restricted to specific fields of science and, hence, the narrow conception of scientism is not a fruitful way to go forward. Instead, one should focus on expanding the understanding of scientism inside the boundaries of the broad-weak variety. Science does not need philosophers to draw boundaries from the sidelines.

The primary reasons to be wary of Narrow-Weak Scientism are the odd consequences that follow from its endorsement. The idea that the natural sciences are always the best for producing knowledge would mean that the methods of the natural sciences should always be preferred over other methods. Now, it is rather evident that there is some knowledge to be had, for instance, of historical subjects, and, currently, the methods used for obtaining this knowledge are those pertaining to the academic field of history—such as cross-examining historical archives to see whether there are multiple independent references to the same event, source criticism, and so on. A proponent of Narrow-Weak Scientism needs to argue either that historical knowledge does not exist or that such knowledge would be easier to achieve or qualitatively better (whatever the standards for that may be) if one would use the methodology of natural sciences, such as physics or chemistry. Both options seem implausible. First, we clearly can and *do* possess adequately justified knowledge of the human past. Second, the proponent of Narrow-Weak Scientism would have to demonstrate how the natural sciences should be applied in studying, say, cultural history. Current historians and archaeologists already use methods and knowledge from the natural sciences when appropriate, for example, in dating artifacts. If the methodology of the natural sciences is superior in all situations, why are historians not already using them across the board? The burden of proof appears to lie on the shoulders of those who support Narrow Scientism.

A natural objection to the above would be the following: accept the fact that the natural sciences cannot produce all imaginable knowledge but argue that whenever they do produce knowledge, they do a better job than other academic fields. The problem with this argument is that it has no practical consequences whatsoever. For cases where natural sciences cannot produce knowledge, it does not matter whether that nonexistent knowledge could be imagined to be better than something else. Indeed, it does not matter whether one is fond of the results of physics than those of history if said fields do not overlap with the subjects their methods can examine.[3]

Narrow-weak forms of scientism have their problems, but we think there is a way to defend the broad-weak forms of scientism. Before turning to our defense of this type of scientism, we will consider Mizrahi's recent work on the topic.

3. MIZRAHI'S DEFENSE OF SCIENTISM AND ITS PROBLEMS

The main argument for scientism has always been the success of science: scientific methods are superior compared to others because science is a success story. While it is debatable what exactly is meant by science being "successful," it is usually undisputed that there is at least some sense in which science is successful. However, a proponent of Strong Scientism must not only show that science is successful but also that no other proper source of knowledge exists or, at the very least, is known to exist. This is a tall order given the sheer amount of different epistemic practices to go through. The odds are thus stacked heavily against a proponent of Strong Scientism. For Weak Scientism, however, the situation is not as dire since it is enough that no other source of knowledge is known to be as good as science. This opens a possibility for a quantitative defense for scientism.

When one states that scientific knowledge is the best knowledge to be had, one can mean two different things, as Mizrahi (2017, 357) points out. Better/ best can be cashed out in terms of *qualitatively* better/best or *quantitatively* better/best. His defense of the quantitative superiority of science rests on a couple of assumptions and empirical methods from information science. His first assumption is that since "the goal of any scholar or academic researcher [. . .] is to publish, it is clear that scientific disciplines are better at attaining this goal" (id, 357). And indeed, the data Mizrahi shows is a clear win for the natural and social sciences (id., 357 fig. 2). His second assumption is that impact values reflect the quality of research. So, if the sciences publish more papers with higher impact factors, then science clearly wins out in quantity. And, again, the data are on Mizrahi's side, as more papers in academic fields outside of the natural and social sciences—that is, the humanities—are left without even a single citation (id., 358 fig. 3). But should these facts worry the opponents of Weak Scientism? And should they give solace to its proponents?

The opponent of scientism might retort that there are fewer resources allocated to her field, that there is a social pressure to gainsay her results, that there are differences in publishing culture, and so on, as reasons for both the lower output of papers and for the lower citations. On top of this, the very use of impact factors to evaluate the quality of work is, at best, questionable (Larivière et al. 2016; Paulus et al. 2018). Moreover, would a proponent of scientism be satisfied with this quantitative argument? After all, it was not the worry that the humanities produce fewer papers that led the proponent of scientism to her views. Instead, the worry was that the methods currently employed, say in analytic philosophy, are a poor fit for their goals (see, e.g., Ladyman et al. 2007, chap. 1).

To hammer the point home, what if a field is like a history, where an emphasis is still put on producing books rather than articles? Why then look at the number of publications, as for such a field output would be *expected* to be low. Further, the low number of citations might be caused by the subject under study being rather a niche, say the production and consumption of rye in northern Slavonia in the latter half of the eighteenth century. After all, other academics might lose interest in particle physics, and still, this would not make particle physics any worse as a science.

Both of Mizrahi's assumptions are questionable. Is the goal of a scholar to publish as much as she can? Given the worries about impact factors, does the impact factor of a paper track its quality? So, equating "better" with "more publications with more citations" seems a non-starter.

What is more promising is equating "best/better" with quality rather than quantity. Here Mizrahi equates "best/better" with *success*. He singles out three types of epistemic success: explanatory, instrumental, and predictive success. Then using two cases, the theory of general relativity from physics and external world realism from philosophy, he evaluates whether scientific theories have a better track record than non-scientific ones. Now, general relativity clearly enjoys all three kinds of epistemic success: it *explains* gravity, *allows for* accurate satellite positioning systems, and *predicts* gravitational redshift, among other things. Things look far bleaker for realism about the external world. It does not explain why we have sensory experiences, as a brain-in-a-vat might as well have those. Instrumentally, it does not allow for any interventions or manipulations. Likewise, there are no novel predictions that can be made from external world realism. Facts like our possibility of sensory experiences would be an *explanandum* of the theory, not its prediction. From these considerations, Mizrahi concludes that science (the theory of relativity) wins out[4] (Mizrahi 2017, 358–59). Here Mizrahi is still keeping his analysis "externalistic"—he is dealing with the *external* criteria for success. Instead of trying to find external markers for the superiority of science, one can try to focus on what makes science a better source of knowledge than other epistemic practices. Science could be epistemically superior because of the way it is done. In other words, science would be successful because of its methodology.

However, the connection between the success of science and its methodology is not straightforward. Science is performed by people, and thus many social and historical factors affect how science is done. Furthermore, the notion of methodology as "how something is done" is very general and vague. Consequently, the connection between the success and methods of science is a somewhat muddled affair. There are, however, ways around this.

4. SCIENTISM: A CONSEQUENCE OF EPISTEMIC OPPORTUNISM

One avenue for a defense of Weak Scientism is to appeal to general methodological principles. One such principle is *epistemic opportunism* (Hietanen et al. 2020). An epistemically opportunist practice adopts any method that *demonstrably works*.[5] Hence, if science is epistemically opportunist, it will always end up being the most, or even the only, successful epistemic endeavor. This is so because it contains, or at least it should contain, all demonstrably working epistemic practices. Here, we have a strong argument for scientism: science is the best (or only) source of evidence, epistemic justification, knowledge, and so forth, since it includes all demonstrably working epistemic practices.[6]

Still, one might wonder whether epistemic opportunism will trivialize scientism. Appealing to it appears to be a victory by definition because the proponent of scientism seems to simply define science as a collection of all epistemic practices that are currently known to be successful. However, one should note that the central premise of the argument can be read in two different ways. It can be understood as a descriptive claim which states that any given science actually is epistemically opportunistic. Alternatively, it can be taken as a normative claim which says that sciences *should* be epistemically opportunistic. The descriptive claim depends on empirical factors, and this makes it non-vacuous. The normative claim, in turn, can be defended on pragmatic grounds by demanding an epistemology that works in practice.

Furthermore, a working epistemology has to be demonstrable. This enables the proponent of scientism to reject epistemic practices such as appealing to pure intuition—that is, to an intuition which one cannot check with independent evidence—or to a *sensus divinitatis*, or something similar.[7] Moreover, the proponent of scientism can, of course, reject all those epistemic practices that demonstrably do not work better than mere chance. To sum up, there are two kinds of epistemic practices, not to mention theories, that the variety of scientism outlined here rejects:

(1) Epistemic practices and theories that are unevaluable. A practice or a theory is unevaluable when it is impossible to demonstrate how well it works.
(2) Epistemic practices and theories that are evaluable but do not work better than random chance.

By a theory "working," we mean that its checkable implications are correct.[8]

Yet, it might feel like this is giving too much leeway. What if, say, reading tea leaves or homeopathy would turn out to be epistemically successful practices? Would they then count as sciences? This seems to go against what we take science to be. Moreover, accepting practices like these as sciences would appear to be a radical departure from current practice, and it would change our scientific worldview drastically (Boudry 2020). Such intuitions are understandable, but there are a few important caveats here.

One of the main reasons for thinking that, as an example, homeopathy is not, and should not be, a part of science is precisely its poor epistemic track record. Indeed, one should not merely assume that some epistemic endeavor is either problematic or praiseworthy. This is an empirical question and not an armchair issue. If homeopathy would have always worked without problems, it would most likely not feel odd to call it a proper part of medical science. To be sure, scientism can demarcate science from nonscience only when the nonsciences actually differ from the sciences. Therefore, if some are conjuring up thought experiments where pseudosciences and other epistemically dubious enterprises work, they also need to bracket, at least for the moment being, their knowledge of the fact that the pseudosciences in question do not actually work. Otherwise, they might end up imagining that pseudosciences are not pseudosciences and then insisting that they should not be counted as sciences because they are pseudosciences. This is, in part, why such thought experiments and intuitions are deceiving. The strong and very sensible intuition that astrology, homeopathy, and tasseography do not belong among the sciences is caused precisely by their continuous failures and discrepancies with proper successful science.

We can also turn the thought experiment around: If we have demonstrably successful epistemic practices, then what grounds do we have for not accepting them into science? Should scientists be forbidden from using some methods that work but not others? If so, on what basis? Should we make a separate class for epistemic practices that work but are not to be included in science? For what purpose would we need such a separation? If we make this distinction, would we have then abandoned the central reason for scientism, namely, the success of science? This path is not an easy one, and taking it requires more than the mere desire to make demarcations.

But let us return to pseudoscience for a moment. As we already argued, there are two types of theories and practices that the proponents of scientism can reject: those that are unevaluable and those that are evaluable but do not work. Pseudosciences, like homeopathy, are guilty of at least one of these two sins. In fact, pseudoscientists frequently commit both of the mentioned failures. Relatively often, some parts of a pseudoscientific doctrine are unevaluable, and others demonstrably incorrect or methodologically unsuccessful.

However, many pseudoscientific doctrines start as something that one can evaluate. For instance, astrology and homeopathy have been evaluated and shown to be erroneous (see, e.g., Hartmann et al. 2006; Grimes 2012). But, of course, that has not secured their demise. Besides ignoring countering evidence, pseudoscientists are also masters at coming up with ad hoc hypotheses that can salvage their theories from rebuttals. Now, unlike a naive Popperian might insist, the use of ad hoc hypotheses to rescue theories from anomalous or otherwise undesired results is not necessarily wrong. Remember that

Neptune was postulated initially as an ad hoc hypothesis to save Newtonian mechanics, and no one thought anything of it. This is because it was possible to check whether there actually was a planet of the required size in the predicted orbit. And as Greg Bamford (1999) has wonderfully illustrated, the problem with ad hoc hypotheses is not that they are ad hoc but that they are often unevaluable. One of the fouls that pseudoscientists perpetrate is that they are ready to settle for ad hoc hypotheses that are not independently testable—in principle or practice. (See also Kitcher 1984, 46–9.)

Besides excluding pseudosciences and superstitious practices, like divination, scientism is an informative position also in other ways. As an example, the proponents of scientism can maintain a difference between science and common sense, even though Quine's (1976, 229) thesis that "[s]cience is not a substitute for common sense but an extension of it" is a sensible one. This is because science can overcome shortcomings of individual epistemology by utilizing social epistemology. A fair amount of evidence supports the idea that human reason is limited by cognitive biases or bounded rationality (see, e.g., Kahneman and Tversky 1996; Gigerenzer and Selten 2002). The scientific community's rigorous error-checking restricts the effects that personal cognitive limitations of scientists would otherwise have on their research. This, among other things, makes scientific knowledge production superior compared to everyday thinking, even if science is based on the more trustworthy parts of common sense.

To conclude, scientism based on epistemic opportunism does not turn everything into science and, thus, it is an informative position.

5. WHAT EPISTEMIC OPPORTUNISM REQUIRES?

However, there is another problem with defending scientism through epistemic opportunism. Epistemic opportunism was defined via appealing to something that demonstrably works. But what does it mean for something to "demonstrably work"? If this notion is left vague, the condition becomes trivial, and scientism loses its punch. We need to impose some criteria for something to work demonstrably, but what criteria should they be?

Too lenient conditions can end up including everything. There are senses in which, say, astrology works. Astrologists can come up with claims that can be interpreted to be correct (such as "something surprising happens to you"), and there is a sense in which those are epistemic claims (they are about something happening in the world). So, in a very loose sense, astrologists can demonstrate how their methods get things right. By epistemic opportunism, they should then be included in science if this reasoning is followed.

On the other hand, too stringent conditions can end up making scientism based on epistemic opportunism untenable. If we, for example, interpret that "demonstrably works" demands that the method is known beforehand to provide correct results for each particular case, then we could only sanction methods that provide epistemic claims we already know. There would be no point in such methods since they could not provide any new information. Such extreme locality would make methodology redundant and any methodological scientism empty. Demanding that a method can be demonstrated to work only by, say, numerical means would also leave out many epistemically useful methodologies, such as many methods of proof used in mathematics that do not require the use of numbers. This would make refuting scientism a trivial matter already on purely pragmatic grounds.

We need to step carefully to avoid such pitfalls. An initial safe step could be to demand that *whatever the set criteria for evaluating whether something works are, the criteria should be presented for the evaluation to take place.* Working is always relative to some *criteria for* working. So, for example, the astrologer must be able to make clear what their prediction "working" would entail irrespective of what actually happens—for the prediction to be evaluable.

The second step could be to demand that *the presentation of those criteria need to be such that one can interpret them in a non-biased manner.* Or, at least, in a way that tries to avoid bias. This requires, at minimum, that the *criteria are given in an intersubjectively accessible way.* There is no way to show that all possible sources of bias are eliminated, but we can nevertheless try to eliminate known sources of bias.

A third step could be about avoiding the problem with the extreme locality. We need to allow methods to be evaluable for some general class of cases. Since evaluations are individual, such more general evaluations need to depend on how the cases in the more general class relate to each other. So, how well a method tested for case X will work for case Y will depend on how X and Y are related. By testing the method in different related cases, we can get some basis for non-local claims about some methodology working. However, this requires that the way the cases are related is presented so that one can evaluate those differences. That is, *all the differing cases a method is used for need to be specified in an evaluable manner.* We need to be able to evaluate in advance in which cases the method should work based on previous applications before we apply it to some new case (see also Turunen 2020, chap. 3).

We can take the fourth step by focusing on the notion of "epistemic." Epistemic opportunism only relates to epistemic methods, but what makes a method epistemic? Well, an epistemic method can provide epistemic claims, but what are epistemic claims? While one can probably say a great deal to answer that question, it seems rather innocuous to take epistemic claims as claims that are either correct or incorrect and that they are about something. From the perspective of evaluating whether a method epistemically works, this seems to imply that we *must be able to evaluate when the claim in question is correct and when it is incorrect.* A method epistemically working would then just be about comparing what according to the method would be correct and what is found to be correct. If that comparison is made in an evaluable way, we can say that the method was demonstrated to (epistemically) work.

With these initial steps, we have gotten some way of clarifying what epistemic opportunism could entail. However, further steps might be needed, and further analysis will hopefully let us find out.

With a defense for Weak Scientism at hand, we can now consider the question of a possible methodological divide within philosophy. Is the battle over the soul of philosophy fundamentally about epistemic opportunism? Is philosophy epistemically opportunistic? Should it be?

6. THAT SOUL OF PHILOSOPHY

Those who have been accused of or even explicitly embraced scientism have often focused their critique on the methods used in philosophy.[9] In particular, those who oppose speculative methods favoring more scientific ones have usually found their critique on the evaluability of speculative methodology. In other words, they take speculative philosophy to be either unevaluable or evaluable but faulty.

For example, James Ladyman, Don Ross, and David Spurrett have criticized analytic metaphysics—which they have pejoratively dubbed "neo-scholastic metaphysics"—for using intuitions as means of justification (Ladyman et al. 2007, vii). Among the problems of intuitions that they bring up are that individuals do not share the same intuitions, so all intuitions cannot be correct, intuitions have developed for cultural and biological purposes and not for discovering metaphysical truths, and scientific findings are often unintuitive (id., 10–13). Thus, according to Ladyman, Ross, and Spurrett, intuitions can be evaluated and proven unreliable. Similarly, Alexander Rosenberg has claimed that science has shown introspection to be "completely untrustworthy as a source of information about the mind and how it works" (Rosenberg 2011, 147). Patricia Churchland, in turn, has disapproved of thought experiments used in traditional analytic philosophy because they typically are "unconstrained, poorly defined, and impossible to evaluate" (Churchland 2002, 265).

How should we react to these criticisms, then? After all, the requirement for intersubjective evaluability, a prerequisite for epistemic opportunism, seems to be quite modest. It is somewhat concerning if such methodological commitments are what the debate on the heart and soul of philosophy is about—as the case indeed seems to be. Yet, the critics of scientism can still be wary. "Granted, epistemic opportunism and epistemic evaluability are not so terrible. But surely," they might insist, "accepting such moderate meta-methodological principles does not lead to full-blown scientism!" Well, maybe not. But then again, perhaps the germ of scientism is already growing within the soul of philosophy? Perhaps, no one was just careful enough to pay attention to it. We hope that the soul of philosophy is not so frail as to fear even the mere possibility of a refutation of one's cherished ideas.

7. CONCLUSIONS

The debate on scientism has shifted gears. Instead of taking scientism as something to be avoided by definition, the debate now considers various forms of scientism and their entailments. In particular, the shift to weak forms of scientism has enabled novel defenses and helped refine what the scientism debate is all about. Part of this is, of course, the question of the nature or soul of philosophy that Mizrahi (2019) has brought forward. Although Mizrahi presents quotes from several philosophers that point to this direction, it is not clear that the debate over scientism is a battle over the soul of philosophy unless one takes the *methods* of philosophy to be coextensive with its soul. In any case, Mizrahi's empirical work, due to its deficiencies, does not support the claim.

In this chapter, we have considered how Weak Scientism can be divided into broad and narrow varieties and argued that the narrow kind faces problems. We paid particular attention to how Mizrahi presented his quantitative defense of scientism and to the flaws with such an approach. We concluded that his methods do not, unfortunately, cut the mustard. If they did, we would appropriate them without hesitation in the spirit of epistemic opportunism. But as things stand, better science—or *Wissenschaft*—is needed. Also, we found Mizrahi's qualitative arguments for scientism lacking.

Not content, we proposed a different approach for defending scientism: appealing to epistemic opportunism. According to epistemic opportunism, an epistemic practice should adopt all methods that demonstrably work. We showed that scientism can be an informative view which does not turn everything into science. It is capable of excluding unevaluable and unworking methods and theories. This includes pseudoscientific, superstitious, and some commonsense beliefs and practices.

Finally, we considered what further assumptions are needed for a non-trivial view on scientism based on epistemic opportunism. We came up with four such assumptions. First, we noted that, whatever the set criteria are for evaluating whether something works, they should be presented for the evaluation to take place. Second, the presentation of those criteria needs to be such that one can interpret them in a non-biased way. This means, in particular, that the criteria have to be given in an intersubjectively accessible way. The third remark related to the possibility of having general knowledge and stated that in all the differing cases, that a method is used for, need to be specified in an evaluable manner. This was so that we would be able to say in advance to which cases the method should work based on previous applications before applying it to a new case. The final remark concerned the epistemic aspect of epistemic opportunism. It stated that we must be able to evaluate when a claim in question is correct and when it is incorrect. These meta-methodological remarks are preliminary but are nevertheless nontrivial. Considerations of Weak Scientism have thus led to meta-methodological considerations. The close connection between philosophy and the scientism debate makes them metaphilosophical.

Epistemic opportunism takes us away from mere demarcation between science and nonscience to methodology. In fact, epistemic opportunism turns such issues of demarcation into methodological ones. The interesting question is what constitutes good methods for producing knowledge, and not under what categories various fields utilizing those methods happen to be organized.

The general lesson to be learned here is that one ought not to think of science as merely a collection of current scientific knowledge and the accepted methods for producing it—though, of course, it is also that. At the core of science lies the attitude: "Let us see if this works." From this, all the rest follows. In a word, science is first and foremost based on the meta-method of epistemic evaluability.

As for the soul of philosophy, we noted that debate seems to be about the evaluability of philosophical methods. The criticized methods are often not evaluable or have been demonstrated not to work. Yet, such methods are fiercely defended despite the minimal nature of the requirement for evaluability. This raises the question: does philosophy really need methods whose epistemic merits cannot be evaluated, or would it simply do better without them?

NOTES

1. It is also worth mentioning that many prominent critics of scientism appear to have theistic commitments (see, e.g., McGrath 2011; Peels 2017; Plantinga 2018; de Ridder 2014; Scruton 2015; Smith 2015; Stenmark 2001; Woudenberg 2018).

2. In Hietanen et al. (2020), we classified Mizrahi, perhaps a bit uncharitably, as a proponent of the narrow variant.

3. There is also the case of James Ladyman's weaker-than-weak scientism (Hietanen et al. 2020, 528fn10). This variety of scientism states that if something can be studied at all, it can be studied through natural science—but this does not necessarily mean that natural science offers the best methods for studying the issue in question. Such a position does not seem to bear any normative consequences regarding how science should be practised.

4. Notice also that Mizrahi uses a rather limited basis for induction: only two cases! Perhaps neither theory is a good representative?

5. Epistemic opportunism should not be confused with Feyerabend's (1993) methodological anarchism. The motto: "anything that *demonstrably works* goes," does not imply that "*anything* goes." For instance, Massimo Pigliucci has erroneously claimed that, according to epistemic opportunism, "science does not follow any methodology" (Pigliucci 2020). There does not appear to be a single correct method of science, but the sciences do share several general methodological or meta-methodological principles. These common preconditions of science already suffice to distinguish opportunism from anarchism. We will discuss some of the principles in more detail in this section and the next one.

6. Here science is just taken as any epistemic practice. One can impose further restrictions on what counts as science, such as having a descriptive function. However, a proponent of scientism should be careful since all additional restrictions need to be tied with the epistemic success of science, or the argument from epistemic opportunism falls flat.

7. This does not imply that epistemic practices or belief systems of, for instance, indigenous cultures should be summarily dismissed. They are perfectly valid as methods or theories as long as they demonstrably work. (For more on non-western science and indigenous knowledge, see, e.g., Harding 2011, part II.)

8. There is a consistent way of analyzing and evaluating such implications that require very few commitments. For a detailed explication, see Turunen (2020).

9. In section 1, "Scientism and the Soul of Philosophy," we considered possible causes for the increase in publications on scientism. There we suggested that the phenomenon could be explained, in part, as a critical reaction to the New Atheist movement. Here, on the other hand, we are saying that philosophers adopt scientism due to methodological reasons. These two need not be in tension since much of the recent literature deals with criticisms of scientism and not with the motivations for scientism.

REFERENCES

Bamford, Greg. 1999. "What is the problem of *ad hoc* hypotheses?" *Science and Education* 8 (4): 375–386. doi: 10.1023/A:1008633808051.

Boudry, Maarten. 2020. "Scientism schmientism! Why there are no other ways of knowing apart from science (broadly construed)." *Blog of the APA*, October 8,

2020. https://blog.apaonline.org/2020/10/08/scientism-schmientism-why-there-are
-no-other-ways-of-knowing-apart-from-science-broadly-construed/.

Churchland, Patricia Smith. 2002. *Brain-Wise: Studies in Neurophilosophy.* Cambridge, MA: The MIT Press.

Dawkins, Richard. 2006. *The God Delusion.* London: Bantam Books.

Feyerabend, Paul. 1993. *Against Method: Outline of an Anarchistic Theory of Knowledge.* 1975. London: Verso.

Gigerenzer, Gerd, and Reinhard Selten. Eds. 2002. *Bounded Rationality: The Adaptive Toolbox.* Cambridge, MA: The MIT Press.

Grimes, David Robert. 2012. "Proposed mechanisms for homeopathy are physically impossible." *Focus on Alternative and Complementary Therapies* 17 (3): 149–155. doi: 10.1111/j.2042-7166.2012.01162.x.

Harding, Sandra. Ed. 2011. *The Postcolonial Science and Technology Studies Reader.* Durham: Duke University Press.

Harris, Sam. 2004. *The End of Faith: Religion, Terror, and the Future of Reason.* New York: W. W. Norton.

Hartmann, Peter, Martin Reuter, and Helmuth Nyborg. 2006. "The relationship between date of birth and individual differences in personality and general intelligence: A large-scale study." *Personality and Individual Differences* 40 (7): 1349–1362. doi: 10.1016/j.paid.2005.11.017.

Hietanen, Johan, Petri Turunen, Ilmari Hirvonen, Janne Karisto, Ilkka Pättiniemi, and Henrik Saarinen. 2020. "How *not* to criticise scientism." *Metaphilosophy* 51 (4): 522–547. doi: 10.1111/meta.12443.

Kahneman, Daniel, and Amos Tversky. 1996. "On the reality of cognitive illusions." *Psychological Review* 103 (3): 582–591. doi: 10.1037/0033-295x.103.3.582.

Kitcher, Philip. 1998. *Abusing Science: The Case Against Creationism.* 1982. Cambridge, MA: The MIT Press.

Knobe, Joshua and Shaun Nichols. 2008. "An experimental philosophy manifesto." In *Experimental Philosophy*, edited by Joshua Knobe and Shaun Nichols, 3–14. Oxford: Oxford University Press.

Ladyman, James, Don Ross, David Spurrett, and John Collier. 2007. *Every Thing Must Go: Metaphysics Naturalized.* Oxford: Oxford University Press.

Larivière, Vincent, Véronique Kiermer, Catriona J. MacCallum, Marcia McNutt, Mark Patterson, Bernd Pulverer, Sowmya Swaminathan, Stuart Taylor, and Stephen Curry. 2016. "A simple proposal for the publication of journal citation distributions." *bioRxiv* preprint. doi: 10.1101/062109.

McGrath, Alister. 2011. *Why God Won't Go Away: Engaging with the New Atheism.* London: Society for Promoting Christian Knowledge.

Mizrahi, Moti. 2017. "What's so bad about scientism?" *Social Epistemology* 31 (4): 351–367. doi: 10.1080/02691728.2017.1297505.

———. 2019. "The scientism debate: A battle for the soul of philosophy?" *Social Epistemology Review and Reply Collective* 8 (9): 1–13.

Paulus, Frieder M., Nicole Cruz, and Sören Krach. 2018. "The impact factor fallacy." *Frontiers in Psychology* 9: 1487. doi: 10.3389/fpsyg.2018.01487.

Peels, Rik. 2017. "The fundamental argument against scientism." In *Science Unlimited? The Challenges of Scientism*, edited by Maarten Boudry and Massimo Pigliucci, 165–84. Chicago: University of Chicago Press.

Pigliucci, Massimo. 2013. "New atheism and the scientistic turn in the atheism movement." *Midwest Studies in Philosophy* 37 (1): 142–153. doi: 10.1111/misp.12006.

Pigliucci, Massimo. 2020. "No matter how you put it, scientism is still a bad idea." *Blog of the APA*, October 22, 2020. https://blog.apaonline.org/2020/10/22/no-matter-how-you-put-it-scientism-is-still-a-bad-idea/.

Plantinga, Alvin. 2018. "Scientism: Who needs it?" In *Scientism: Prospects and Problems*, edited by Jeroen de Ridder, Rik Peels, and René van Woudenberg, 220–232. Oxford: Oxford University Press.

Quine, W. V. O. 1976. "The scope and language of science." In *The Ways of Paradox and Other Essays*, 228–45. 1957. Cambridge, MA: Harvard University Press.

de Ridder, Jeroen. 2014. "Science and scientism in popular science writing." *Social Epistemology Review and Reply Collective* 3 (12): 23–39.

———. 2019. "Against empirical-ish philosophy: Reply to Mizrahi." *Social Epistemology Review and Reply Collective* 8 (12): 8–12.

Rosenberg, Alexander. 2012. *The Atheist's Guide to Reality: Enjoying Life Without Illusions*. 2011. New York City: W. W. Norton.

Scruton, Roger. 2015. "Scientism and the humanities." In *Scientism: The New Orthodoxy*, edited by Richard N. Williams and Daniel N. Robinson, 131–146. London: Bloomsbury Academic.

Smith, James K. A. 2015. "Science as cultural performance: Leveling the playing field in the theology and science conversation." In *Scientism: The New Orthodoxy*, edited by Richard N. Williams and Daniel N. Robinson, 177–191. London: Bloomsbury Academic.

Stenger, Victor J. 2009. *The New Atheism: Taking a Stand for Science and Reason*. New York: Prometheus Book.

Stenmark, Mikael. 2001. *Scientism: Science, Ethics and Religion*. Farnham: Ashgate.

Turunen, Petri. 2020. "Constrained realism: Ontological implications of epistemic access." PhD diss., University of Helsinki.

van Woudenberg, René. 2018. "An epistemological critique of scientism." In *Scientism: Prospects and Problems*, edited by Jeroen de Ridder, Rik Peels, and René van Woudenberg, 167–189. Oxford: Oxford University Press.

Chapter 6

Philosophy of Science in Practice and Weak Scientism Together Apart

Luana Poliseli and Federica Russo

The term "scientism" has not attracted consensus about its meaning or about its scope of application. In this paper, we consider Mizrahi's suggestion to distinguish "Strong" and "Weak" scientism, and the consequences this distinction may have for philosophical methodology. While we side with Mizrahi that his definitions help advance the debate, by avoiding verbal dispute and focusing on questions of method, we also have concerns about his proposal as it defends a hierarchy of knowledge production. Mizrahi's position is that Weak Scientism should be adopted, stating that "of all the knowledge we have, scientific knowledge is the *best* knowledge." This version of scientism, however, has consequences for philosophical methodology. In particular, if one conceives of philosophy as an a priori discipline and holds Weak Scientism, the introduction of empirical methods in philosophy may threaten its very essence or soul. In this chapter, we will defend the move to adopt empirical methods in philosophy and argue that, rather than threatening its essence or soul, these methods put philosophy in a better position to contribute to knowledge production, an endeavor shared with the sciences, and in a very interdisciplinary spirit. Our point of disagreement with Mizrahi is that we should avoid any hierarchy of knowledge and instead focus on what each perspective—scientific, philosophical, historical, or other—can contribute to understanding phenomena.

1. INTRODUCTION

Scientism is commonly associated with the idea that scientific method has almost no limits and can successfully be applied to all domains, providing explanation for everything in the world (Schults 2002). Despite its wide range of definitions,[1] the term is often used pejoratively by some philosophers (Mizrahi 2017a) as a response to scientists who assume that philosophy can be rather sterile (Pigliucci 2002, 115) or a waste of time (Weinberg 1992), and therefore is no longer useful (Hawking and Mlodinow 2010). Defenders of scientism have adopted it half-wittingly without a very well-articulated argumentation in its favor, as pointed out by Woudenberg et al. (2018, 2): "it is a view that appears to be more 'in the air' than pinned down on paper as a philosophical position." In order to properly assess this debate and avoid verbal disputes, Mizrahi (2017a) develops an epistemological thesis of scientism that distinguishes between *Strong Scientism* and *Weak Scientism.*

According to this account, *Strong Scientism* assumes that "of all the knowledge we have, scientific knowledge is the *only* 'real knowledge'" (Mizrahi 2017a, 353), while *Weak Scientism* assumes that "of all the knowledge we have, scientific knowledge is the *best* knowledge" (Mizrahi 2017a, 354). The one addressed by Mizrahi as having a significant impact on philosophy is the latter. His argument suggests that scientific knowledge is quantitatively and qualitatively better than non-scientific knowledge because it produces more research outputs and impact, as it is more successful in its instruments and in developing predictive explanations (Mizrahi 2017b, 2018). That being established, he also suggests "that the introduction of methods from data science into logic (Mizrahi 2019) and philosophy (Mizrahi 2018b, 48) might bring to logic and philosophy the sort of success enjoyed by the sciences" (Mizrahi 2019a, 10).

More recently, in "The scientism debate: a battle for the soul of philosophy," Mizrahi (2019b) argues that the question whether philosophy can reach the "success enjoyed by the sciences" gravitates around two fundamental contexts: philosophy as a field of study and philosophy as a field of inquiry. He develops a bibliographic survey and regression correlation analysis with data mined from JSTOR Data for Research, in order to test the following hypotheses:

- H1: many philosophers find scientism intimidating because they see it as a threat to the future of philosophy as a major in college and universities. If this hypothesis is true, philosophers would feel endangered because students would choose STEM majors instead of philosophy.
- H2: many philosophers find scientism threatening because they see it as a danger to the soul or essence of philosophy as an a priori discipline, which is in essence how Analytic Philosophy is typically conceived. If this hypothesis is true, then philosophers would feel threatened when traditional methods of philosophical investigation lose ground to empirical methods.

His results show a very weak positive correlation for H1, meaning that there is a "weak positive correlation between the number of Philosophy and Religious Studies Bachelor's degrees conferred by postsecondary institutions in the United States and the number of publications in the Philosophy subject category on JSTOR that contain the term 'scientism'" (Mizrahi 2019b, 8). While for H2, there is a "correlation between the number of philosophy publications that contain the term 'scientism' and those that contain the phrase 'experimental philosophy'" (Mizrahi 2019b, 9). The results point to the existence of a methodological shift suggesting that philosophy is stepping "away from purely *a priori* methods of investigation (so-called 'armchair philosophy') toward *a posteriori* or empirical method of investigation (so-called 'experimental philosophy')" (Mizrahi 2019b, 10). On this account, philosophy would be subjected to the empirical methods of observation and experimentation that are widely used in science, complying with what is called *Weak Scientism*. However, it is worthy to highlight that even Mizrahi (2019b, 10) acknowledges that these results cannot be explicitly interpreted as good or bad for philosophy, even though he does not elaborate on why it is so.

In our view, *Strong* and *Weak Scientism* theses do possess stronger epistemological appeal than previous definitions offered in the literature as they avoid verbal disputes, persuasive and question begging definitions (see Mizrahi 2017a). We also side with Mizrahi that *Weak Scientism* is, out of the two, the most interesting option to consider. However, we will show that *Weak Scientism*, as presented by Mizrahi, might be misleading because

(i) it obscures the fact that "empirical philosophy" is in fact interdisciplinary (as argued by Bishop 2019) and inherently methodologically diverse (our specific point); and
(ii) it introduces a hierarchy of knowledge production (see Mizrahi 2018b, 2018c responses to Wills 2018a, 2018b) reminiscent of positivism and neo-positivism.

To develop our arguments, we will consider Philosophy of Science in Practice (PSP) as a field of inquiry, developed within philosophy of science. Our goal in this chapter is to show that endorsing Weak Scientism *together with* a skeptical attitude toward the adoption of empirical methods in philosophy may undermine flourishing philosophical approaches such as PSP and socially engaged philosophy. This is because *Weak Scientism*, if taken literally and as not including philosophy, is unable to recognize the inherent interdisciplinary character and methodological diversity of PSP investigations. Thus, instead of focusing on whether and how philosophy may be threatened by the adoption of methods, the question to ask is what *else* we can learn from empirical and empirically informed philosophy that armchair methods *alone* cannot achieve. This, we argue, is the value of interdisciplinarity and of methodological diversity of PSP. As a consequence, the goal should not be to establish rigid hierarchies of knowledge as Mizrahi proposes but to figure out which method is best suited to the research questions and goals one sets.

2. WEAK SCIENTISM AND METHODOLOGICAL DIVERSITY

According to Mizrahi, H2—namely that many philosophers find scientism threatening because they see it as a threat to the soul or essence of philosophy as an a priori discipline—is confirmed by the existence of a positive correlation between the number of publications mentioning the terms "experimental philosophy" and "scientism." He says:

> This strong positive correlation between the number of philosophy publications in which the term "scientism" occurs and those in which the phrase "experimental philosophy" occurs is what we would expect to find if H2 were true. As mentioned above, if H2 is true, we would expect philosophers to feel more threatened by scientism when they think that the traditional methods of philosophical investigation (such as the method of case) begin to lose ground to empirical methods of investigation. This positive correlation and the result of a linear regression analysis, which indicates that the number of "experimental philosophy" publications predicts "scientism" publications in Philosophy, suggest a link between the introduction of empirical methods into philosophy and concerns about scientism among philosophers that is worthy of further investigation, or so I think. (Mizrahi 2019b, 9)

In a commentary to Mizrahi, Bishop (2019) points to two issues in this analysis. First, such correlation is not enough to conclude that these publications focus on empirical philosophy and that scientism has positive or negative connotations (also agreed by Mizrahi *ibid*); both conclusions would require a finer-grained examination of those articles. Second, Bishop brings attention to the misleading aspect of calling the adoption of empirical methods in philosophy "scientism," when it should be considered an instance of interdisciplinarity. We broadly agree with Bishop's concerns and therefore will not focus on those aspects, as they are already addressed in his work. Instead, we wish to follow up on Bishop's arguments and point out that the introduction of empirical and scientific methods into philosophical investigations does not just enrich philosophy with interdisciplinarity aspects but also introduces an inherent methodological diversity, which remains implicit when using the notion of "empirical philosophy." Thus, in the next section, we introduce the "practice turn" in philosophy of science to show that methodological diversity, and the adoption of empirical methods, enriches philosophical methodology, rather than threatening its essence, against Mizrahi's H2.

2.1 The Practice Turn in Philosophy of Science

Philosophy of science, as a field of inquiry, focuses on the nature and production of scientific knowledge through ontological, epistemological, methodological, and normative investigations. According to the reconstruction of Hans Radder, until the 1970s, philosophy of science was a relatively small field (with a few journals, special editions, and monographic works), usually characterized as a general philosophy of science, having physics as the discipline that best represents the ideals of unification and reduction—the role model for all sciences (Radder 2012). After this period, the discussions about the limits of unification, together with new interesting foci in the cognitive sciences, pushed toward the development of a philosophy of special sciences (Fodor 1974; Schurz 2014). As a consequence, there was a fragmentation and subsequent specialization: journals focusing on specific scientific disciplines, topics, and methods have increasingly emerged (Radder 2012). This "speciation" of philosophy of science has been visible in the gradual inclusion of philosophy of biology and of the life sciences or philosophy of the social sciences and of economics as legitimate subfields and, more recently, of philosophy and methodology of medicine.

On the one hand, philosophy of science has considerably broadened the scope of questions asked, going beyond the very classic themes of explanations, natural laws, (anti)realism, objectivity, methodology, ethics/value, demarcation and progress across the science, and also developed concepts, theories, and methods that are tailored to specific scientific fields. On the other hand, such fragmentation and subsequent specialization have reduced its impact to the outside sphere of its practitioners, being even called as the "Siberia of Philosophy" (Radder 2012). Despite its theoretical advances regarding the practices of science, traditional philosophy of science has aimed at an account of scientific knowledge in terms of a two-way relationship between world and knowledge (Boon 2017). Such philosophical analyses were mainly based on interpretations of already published theories and results, instead of engaging with real-life practices run by scientists, thus allowing an analysis of the ongoing process of experimentation, data gathering, and so on. In the 1980s, the call for an analysis of science that would put the practice at the center had been advocated not only in philosophy but also, and perhaps primarily, in some fringes of the history of science and in sociology of science. The main idea was that we should focus on the scientific *practice*, in its social, political, material, and psychological dimensions (for a reconstruction of the practice turn, see Soler et al. 2014).

Other arguments in favor of focusing on the distinct practices of science came from post-Kuhnian, naturalists and postmodernists (Radder 2012). Post-Kuhnians defended that science could not be understood only through an armchair perspective (see Bryson 2009), while naturalists questioned the distinction between philosophy and science arguing that to understand scientific activities, such as observation or theoretical reasoning, one needs to engage with cognitive sciences. And lastly, the postmodernists have improved fundamental criticisms about a general epistemology of science, of universal theories of rationality, and understandable scientific methodologies (Radder 2012). Outside the philosophy of science, pragmatists such as Charles Peirce and John Dewey, as well as the later Wittgenstein, have also dedicated attention to search truth and meanings through practices and instruments (see Stern 2003). Continental philosophical traditions have also highlighted the need to consider practices and experiences while they have also rejected positivists' traditions that saw sciences as excessively privileged, having scientific progress as right (Ankeny et al. 2011).

In the practice turn, social and historical studies of science (and technology) focused on scientific practices as products of human activity, while philosophy of science focused on the relations between scientific theories and the world. However, philosophy of science was still isolated from the scientific practices per se and being developed mainly through armchair reflections (Ankeny et al. 2011). Thus, if the goal was to understand and explore the methods and frameworks underlying scientific practices, both scopes were limited, not only because they neglected the perspectives and approaches important for a more complete picture of science, but also because they neglected the *processes*, or practices, that lead to scientific conclusions and scientific products (Ankeny et al. 2011). Echoing such concerns and addressing them, PSP started as a learned society in philosophy and as a professional venue to facilitate and foster debate.

PSP, among other things, aims to account for scientific practices in various fields and addresses questions such as: How is the construction of knowledge for epistemic uses possible? What methods are employed in such constructions? And what do methods and concepts tell us about the world? How are these practices value-laden? PSP aims at an understanding of science that avoids the belief that the objectivity of knowledge can be warranted by an account of knowledge-justification that eliminates the role of scientists, but that also avoids a mere psychological and sociological interpretation of scientists' subjectivity (Boon 2017).

The Society for Philosophy of Science in Practice (SPSP) defines the term "practice" as organized or regulated activities that aim to achieve certain goals (see, e.g., Chang 2014). Thus, any investigation of practices should elucidate what kind of activities are associated with them and required for the generation of knowledge in a given domain. In this sense, PSP has the practice*s* of science as its object of research.

2.2. Philosophy of Science in Practice Is Methodologically Diverse

PSP does not possess any general protocol or any specific methodology to apply in order to achieve its goals. The instruments used to investigate the practices of the sciences come from history, psychology, technology studies, sociology, and so on (Boon 2017). These instruments include, but are not limited to, conceptual analysis, historical reconstruction and contextualization, analysis and consideration of cultural, social, political aspects, discourse analysis, formal methods, or ethnographic approaches. It is important to note that there is no hierarchy in this list of methods or clusters of methods—unlike what is suggested by Weak Scientism. The lack of a general methodological approach does not make PSP more or less valid as a whole and for this reason does not squarely fit with any definition of scientism (weak or strong). The abovementioned methodological instruments, with their diversity, constitute a toolbox to achieve the goal of understanding various aspects of scientific practices. Instead of sticking to any rigid protocol, PSP takes advantage of a set or family of approaches from different fields, configuring what we would like to call *methodological diversity*. The challenge, then, is to map how these methodological processes might happen in this wide range of possibilities.

Attempting to organize such methodological diversity in the context of PSP, John Dupré, at the 3rd Biennial Conference for the European Philosophy of Science Association at the University of Exeter, 2011, has called attention to the duality of approaches and applications depicted by *philosophy of science-in-practice* and *philosophy-of-science in practice* (Boumans and Leonelli 2013). The first one is the philosophy that analyzes science being made, the daily activities associated with scientific research, in this approach, philosophers do not necessarily collaborate with scientists but they can use empirical methods from history or sociology (Kosoloski 2012; Boumans and Leonelli 2013). The second one is the philosophy that engages with the scientific research through the interaction with scientists about philosophical problems and/or common ground issues. This approach does not require that philosophers engage with empirical methods, even though this might occur, the emphasis is to recognize shared problems in philosophy and science (Kosoloski 2012; Boumans and Leonelli 2013). A socially engaged philosophy is also part of such methodological diversity. Interesting work has been developed by initiatives such the Toolbox Project, at the SRPoiSE—Socially

Relevant Philosophy of/in Science and Engineering (http://srpoise.org); a *Socially-Engaged Philosophy* organized by Martin Kush at the University of Vienna; and *The Geography of Philosophy* (www.geographyofphilosophy .com) by Edouard Machery, Stephen Stich, and H. Clark Barret. Recently, Laplane et al. (2019) have also called attention to the important role and impact philosophy can have in science, although some scientists still conceive philosophy as being antagonistic to science. Drawing on three examples from contemporary life sciences (i.e., cognitive sciences, immunology, and stem cell research), these authors show that philosophical contributions to science can be of at least four kinds: clarification of scientific concepts, critical assessment of scientific assumptions and methods, development of new concepts and theories, and fostering dialogue between distinct fields. Laplane et al. (2019) article fosters this debate by arguing that modern science can benefit from philosophy, and that this close integration between the fields can enhance the vitality of science. This specific approach stems from PSP and is called Philosophy *in* Science (Pradeu et al. 2021).

According to Tress et al. (2005), interdisciplinarity can create a unique body of knowledge that merges from disciplinary fields. Elaborating on Tress et al. (2005), Poliseli (2018) also notes that in the interdisciplinary work of PSP, philosophical and scientific knowledge are *co*-produced. In such interplay of empirical and philosophical methods with scientific methods, there is no hierarchy of methods for knowledge production—unlike what is suggested by Weak Scientism and by accredited methodologies such as evidence-based medicine, that are quite reminiscent of positivist approaches to science and knowledge. This can be illustrated through many examples in PSP.

In the following, we select two examples from our own work, because they closely follow the spirit and intentions of PSP to understand and make sense of scientific practices and to actively contribute to advancing the scientific process. While they may seem prima facie very similar methodologically, the first case (but not the second) uses ethnographic methods, and the second (and to a lesser extent the first) uses conceptual analysis (informed by the practice of science) in order to formulate explicit recommendations for public health. In these examples, philosophers can use philosophical tools to tackle scientific problems and therefore produce scientific knowledge (Chang 1999), but can also address philosophical problems using lessons taken from case studies in science (Malaterre et al. 2019).

Our first example is taken from Poliseli (2018), who elaborates an account of how a philosopher can actively contribute to scientific practice and at the same time produce philosophical knowledge. Poliseli (2018) presents a case in which a philosopher together with an ecologist can produce a set of heuristics that is built on a combination of philosophical and ecological knowledge. The combination and synergy of methods from philosophy and ecology were carried out in view of studying mechanistic model-building and assessing theories of scientific understanding. This unique body of heuristics can be analyzed by both scientists and philosophers, through scientific methods (e.g., model-building) and traditional philosophical conceptual analysis (e.g., by engaging with concepts of explanation and understanding) (see Poliseli 2020). Our second example is taken from Kelly and Russo (2017), who aim to develop an account of "mixed mechanisms" in which philosophical analyses of "mechanisms" are combined with social science approaches to health and disease. In the philosophical analysis, they advance the view that a concept of "mixed mechanism" can capture the role of both biological and social factors in the etiology and development of health and disease. The study of mixed mechanisms, however, cannot be reduced to biochemistry but needs social science in a fundamental manner. This combination of methods leads to developing a concept of "lifeworld" that (i) captures the lived experience of individuals and groups, (ii) can be operationalized and studied in quantitative and qualitative studies, and (iii) can be put to use in the context of public health interventions.

While we acknowledge and praise the methodological diversity and the engagement of the practice typical of PSP studies, we should also mention that, as yet, no proper discussion of how methods of the social science can be really combined with typical philosophical methods exists. As a consequence, PSP is still in need of further development, and notably about developing methodological foundations explaining how philosophy can appropriately make use of ethnography and of empirical case studies. For instance, the role of case studies in philosophy (of science) has sparked debate (Chang 2011; Illari and Russo 2014; Mizrahi 2020), but other than mention problematic aspects of this practice in philosophical analysis, no positive solution has been offered, and yet case studies are routinely used in History and Philosophy of Science, PSP, and in other subfields too. This is to say that, while we firmly side with PSP for its interdisciplinary character and its methodological diversity, we also acknowledge that potential problems identified by Mizrahi in applying Weak Scientism to philosophy are not entirely resolved.

Now, returning to Mizrahi's defense of Weak Scientism and how philosophers understand it is as a threat to the soul of philosophy, he says:

> As far as research is concerned, scientism is perceived as a threat to the sort of research that philosophers typically do because it advocates the use of empirical methods of observation, experimentation, and the like, whereas philosophers are typically content with armchair reflections. [. . .] Many philosophers seem to think that scientism poses a threat to them as researchers because it somehow implies that philosophy has no valuable contributions to make to the advancement of knowledge unless it adopts the empirical methods of the sciences.

For this reason, some philosophers find it necessary to defend the traditional methods of philosophy against any attempt to introduce empirical methods into philosophy. (Mizrahi 2019b, 1)

In our view, the abovementioned PSP examples (as well as many others) do not fall under the umbrella of scientism, because their philosophical underpinnings are not dictated by the scientific methods. Instead, philosophical questions motivate the gathering of data of different sorts, to be further substantiated in the analysis of scientific practices. Any armchair reflection could not obtain results at the conceptual, epistemological, methodological, or normative level as is the case in philosophy of science-in-practice or philosophy-of-science in practice. In this sense, the absence of a "coded" methodology opens a multitude of opportunities, precisely for the reason that there is no such thing as a "fixed" or standard univocal way of determining the contents of knowledge production.

In our reading, *Weak Scientism* is accompanied by a hierarchy of methods for knowledge production, and "a priori" methods sit at the very top of a pyramid and "a posteriori" or empirical methods sit at the very bottom of it. Instead, PSP as a field of inquiry takes advantage of methodological diversity, with no strict or rigid hierarchy associated with it. Granted, for any given research question we may come up with an ordered preference of methods, likely to deliver more or less solid results, but such ordered preferences are very local, and in no way universal. This inherent methodological diversity, in turn, calls for practices that are ipso facto collaborative and collegial. As it often happens in PSP circles, scholars from different perspectives come to collaborate on similar problems, in order to produce and generate knowledge that would not otherwise be generated by using a single disciplinary approach (for philosophy of research team collaboration, see Bammer 2013; Andersen 2016; Macleod et al. 2019). The same goes for a philosopher delving into interdisciplinarity practices because they use both traditional philosophical (armchair philosophy) and also some kind of empirical method, or empirically informed approach.

Thus, PSP constitutes an excellent example of how one could hold Weak Scientism, *without* subscribing to any hierarchy of methods for knowledge production; in fact, both philosophical and scientific knowledge are produced in accordance with "traditional" *and* empirically informed philosophical analysis. In a nutshell, there is no better knowledge, only knowledge produced by methods that are more or less suited to explain a certain type of phenomenon or other.

3. CONCLUSION

In this chapter, we have tried to explore some challenges posed by the concept of *Weak Scientism*, in particular as a thesis that may threaten the future of philosophy (of science). We have shown that the introduction of empirical methods in philosophy does not put philosophy under threat of losing its essence of distinct a priori character. Instead, we take the inherent interdisciplinarity and methodological diversity of empirically oriented method as an enrichment of the toolbox of philosophers. In that sense, there is no reason to see philosophy outside the scope of Weak Scientism, or in competition with the sciences. There are flourishing philosophical approaches such as PSP and socially engaged philosophy that are inherently interdisciplinary and methodologically diverse, and their contribution to studying scientific problems ought to fall under a qualified version of *Weak Scientism* too, qua approaches that aim at producing valid knowledge. In particular, our qualification of *Weak Scientism* is that it should not entail a hierarchy of knowledge. As we have tried to defend here, the question is not about which knowledge is best, but which knowledge better explains their targeted goal, using appropriate tools, of any kind.

Our positive suggestion is that, instead of focusing on how philosophy is threatened (or not) by scientific methods, the question should be what *else* we can learn from empirical and empirically informed philosophy that armchair methods *alone* cannot suffice to achieve. This, we argue, is the value of interdisciplinarity and of methodological diversity: not to establish hierarchies, but to figure out which method is best suited to the research questions and goals one sets.

NOTE

1. For others definitions and positions about scientism see Wilson (1999), Olafson (2001), Stenmark (2001), Margolis (2003), Haack ([2007]2015), Olson (2008), Rosenberg (2011), and for a conceptual map see Peels (2018).

REFERENCES

Andersen, Hanne. 2016. "Collaboration, Interdisciplinarity, and the Epistemology of Contemporary Science." *Studies in History and Philosophy of Science Part A* 56: 1–10.

Ankeny, Rachel, Hasok Chang, Marcel Boumans, and Mieke Boon. 2011. "Introduction: Philosophy of Science in Practice." *European Journal for Philosophy of Science* 1 (3): 303–307.

Bammer, Gabriele, Simon Bronitt, L. David Brown, Marcel Bursztyn, Maria Beatriz Maury, Lawrence Cram, Ian Elsum, Holly J. Falk-Krzesinski, Fasihuddin, Howard Gadlin, L. Michelle Bennett, Budi Haryanto, Julie Thompson Klein, Ted Lefroy, Catherine Lyall, M. Duane Nellis, Linda Neuhauser, Deborah O'Connell, Damien

Farine, Michael O'Connor, Michael Dunlop, Michael O'Rourke, Christian Pohl, Merritt Polk, Alison Ritter, Alice Roughley, Michael Smithson, Daniel Walker, Michael Wesley, and Glenn Withers. 2013. *Disciplining Interdisciplinarity: Integration and Implementation Sciences for Researching Complex Real-World Problems*. Canberra: ANU Press.

Bishop, Robert C. 2019. "Scientism or Interdisciplinarity?" *Social Epistemology Review and Reply Collective* 8 (12): 46–49.

Boon, Mieke. 2017. "Philosophy of Science in Practice: A Proposal for Epistemological Constructivism." In *Logic, Methodology and Philosophy of Science*, edited by Hannes Leitgeb, Ilkka Niiniluoto, Päivi Seppälä, and Elliot Sober, 289–310. Proceedings of the 15th International Congress (CLMPS 2015): College Publications.

Boumans, Marcel and Sabina Leonelli. 2013. "Introduction: On the Philosophy of Science in Practice." *Journal of General Philosophy of Science* 44 (2): 259–261.

Chang, Hasok. 1999. "History and Philosophy of Science as a Continuation of Science by Other Means." *Science & Education* 8 (4): 413–425.

Chang, Hasok. 2011. "Beyond Case-Studies: History as Philosophy." In *Integrating History and Philosophy of Science: Problems and Prospects*, edited by Seymour Mauskopf and Tad Schmaltz, 109–124. Dordrecht: Springer.

Chang, Hasok. 2014. "Epistemic Activities and Systems of Practice: Units of Analysis in Philosophy of Science After the Practice Turn." In *Science After the Practice Turn in the Philosophy, History and Social Studies of Science*, edited by Léna Soler, Sjoerd Zwart, Michael Lynch, and Vincent Israel-Jost, 67–79. New York: Routledge.

Fodor, Jerry. 1974. "Special Sciences (Or: The Disunity of Science as a Working Hypothesis)." *Synthese* 28 (2): 97–115.

Haack, Susan. 2007. *Defending Science Within Reason: Between Scientism and Cynicism*. New York: Prometheus Books.

Hawking, Stephen and Leonard Mlodinow. 2010. *The Grand Design*. New York: Bantam Books.

Illari, Phyllis and Federica Russo. 2014. *Causality: Philosophical Theory Meets Scientific Practice*. Oxford: Oxford University Press.

Kelly, M. and Federica Russo. 2017. "Causal Narratives in Public Health: The Difference Between Mechanism of Aetiology and Mechanisms of Prevention in Non-Communicable Diseases." *Sociology of Health & Illness* 40 (1): 82–99.

Kosoloski, Lazlo. 2012. "Philosophy-of-Science in Practice Vs. Philosophy of Science-in-Practice." *SPSP Newsletter* 2 (Autumn): 9–10.

Laplane, Lucie, Paolo Mantovani, Ralph Adolphs, Hasok Chang, Alberto Mantovani, Margaret McFall-Ngai, Carlo Rovelli, Elliott Sober, and Thomas Pradeu. 2019. "Why Science Needs Philosophy." *PNAS* 116 (10): 3948–3952.

MacLeod, Miles, Martina Merz, Uskali Mäki, and Michiru Nagatsu. 2009. "Investigating Interdisciplinary Practice: Methodological Challenges (Introduction)." *Perspectives on Science* 27 (4): 545–552.

Malaterre, Christophe, Chartier Jean-François, and Davide Pulizzotto. 2019. "What is This Thing Called Philosophy of Science? A Computational Topic-Modeling

Perspective, 1934–2015." *Hopos: The Journal of the International Society for the History of Philosophy of Science* 9 (2): 215–249.

Margolis, Joseph. 2003. *The Unravelling of Scientism: American Philosophy at the End of the Twentieth Century*. Ithaca, NY: Cornell University Press.

Mizrahi, Moti. 2017a. "What's so Bad About Scientism?" *Social Epistemology Review and Reply Collective* 31 (4): 351–367.

Mizrahi, Moti. 2017b. "In Defense of a *Weak Scientism*: A Reply to Brown." *Social Epistemology Review and Reply Collective* 6 (11): 9–22.

Mizrahi, Moti. 2018a. "More in Defense of Weak Scientism: Another Reply to Brown." *Social Epistemology Review and Reply Collective* 7 (4): 7–25.

Mizrahi, Moti. 2018b. "Weak Scientism Defended Once More." *Social Epistemology Review and Reply Collective* 7 (6): 41–50.

Mizrahi, Moti. 2018c. "Why Scientific Knowledge Is Still the Best." *Social Epistemology Review and Reply Collective* 7 (9): 18–32.

Mizrahi, Moti. 2019a. "What Isn't Obvious about 'Obvious': A Data-Driven Approach to Philosophy of Logic." In *Advances in Experimental Philosophy of Logic and Mathematics*, edited by Andrew Aberdein and Matthew Inglis, 201–224. London: Bloomsbury.

Mizrahi, Moti. 2019b. "The Scientism Debate: A Battle for the Soul of Philosophy?" *Social Epistemology Review and Reply Collective* 8 (9): 1–13.

Mizrahi, Moti. 2020. "The Case Study Method in Philosophy of Science: An Empirical Study." *Perspectives on Science* 28 (1): 63–88.

Olafson, Frederik A. 2001. *Naturalism and the Human Condition: Against Scientism*. London: Routledge.

Olson, Richard G. 2008. *Science and Scientism in Nineteenth Century Europe*. Chicago, IL: University of Illinois Press.

Peels, Rik. 2018. "A Conceptual Map of Scientism." In *Scientism: Prospects and Problems*, edited by Jeroen de Ridder, Rik Peels, and René van Woudenberg, 28–56. New York: Oxford University Press.

Pigliucci, Massimo. 2002. *Denying Evolution: Creationism, Scientism, and the Nature of Science*. Sunderland, MA: Sinauer Associates.

Poliseli, Luana. 2018. "*When Ecology and Philosophy Meet: Constructing Explanation and Assessing Understanding in Scientific Practice*." Ph.D. Dissertation (History, Philosophy and Science Teaching Program), Federal University of Bahia, Salvador.

Poliseli, Luana. 2020. "Emergence of Scientific Understanding in Real-Time Ecological Research Practice." *History and Philosophy of the Life Science* 42 (4). DOI: 10.1007/s40656-020-00338-7.

Pradeu, Thomas, Mäel Lemoine, Mahdi Khelfaoui, and Yves Gingras. 2021. "Philosophy in Science: Can philosophers of Science Permeate Through Science and Produce Scientific Knowledge." *The British Journal for the Philosophy of Science*. DOI: 10.1086/715518.

Radder Hans. 2012. "What Prospects for a General Philosophy of Science?" *Journal for General Philosophy of Science* 43 (1): 89–92.

Rosenberg, Alex. 2011. *The Atheist's Guide to Reality: Enjoying Life Without Illusions*. New York: W.W. Norton.

Schults, F. Leron. 2002. "Book Review: Scientism: Science, Ethics and Religion by Mikael Stenmark." *Ars Disputandi* 2 (1): 43–46.

Schurz, Gerhard. 2014. *Philosophy of Science: A Unified Approach*. New York: Routledge.

Soler, Léna, Sjoerd Zwart, Michael Lynch, and Vincent Israel-Jost. 2014. *Science After the Practice Turn in the Philosophy, History and Social Studies of Science*. New York: Routledge.

Stenmark, Mikael. 2001. *Scientism: Science, Ethics and Religion*. Aldershot: Ashgate.

Stern, David. 2003. "The Practical Turn." In *The Blackwell Guide to the Philosophy of Social Sciences*, edited by Stephen Turner and Paul A. Roth, 185–206. Oxford: Blackwell Publishing.

Tress, Barbel, Gunther Tress, and Gary Fry. 2005. "Defining Concepts and the Process of Knowledge Production in Integrative Research." In *From Landscape Research to Landscape Planning: Aspects of Integration, Education and Application*, edited by Barber Tress, Gunther Tress, Gary Fry, and P. Opdam, 13–26. Dordrecht: Springer.

Weinberg, Steven. 1992. "Against Philosophy." In *Dreams of a Final Theory: The Scientist's Search for the Ultimate Laws of Nature*, 166–190. New York: Pantheon.

Wills, Bernard. 2018. "On the Limits of Any Scientism." *Social Epistemology Review and Reply Collective* 7 (7): 34–39.

Wills, Bernard. 2018. "Why Mizrahi Needs to Replace Weak Scientism With an Even Weaker Scientism." *Social Epistemology Review and Reply Collective* 7 (5): 18–24.

Wilson, Edward Osborne. 1999. *Consilience: The Unity of Knowledge*. New York: Vintage Books.

Woudenberg, René van, Rik Peels, and Jeroen de Ridder. 2018. "Introduction: Putting Scientism on the Philosophical Agenda." In *Scientism: prospects and problems*, edited by Jeroen de Ridder, Rik Peels, and René van Woudenberg, 1–27. New York: Oxford University Press.

Chapter 7

"Science in the Crosshairs"

Catherine Wilson

To understand and contribute to debates about scientism, we need to start with science. The history of science is one of favor—as evidenced by lucrative prizes, inscriptions, busts and statues, appellations of "genius," the patronage of kings in former times, and staggering research endowments today—but also of fear. Not only skepticism and criticism have stalked scientific practice since its appearance in the ancient world, but also deeply rooted cultural anxieties.

"Science hesitation," to borrow a euphemism, is not new: St. Augustine and many theologians following him condemned curiosity. A human being should meditate on the state of his soul and his relationship to God, not be diverted by the beauty and complexity of nature or questions about how it all came to be. From Plato's attack on the ancient atomists, to the hostility to scientific investigation on the part of the Catholic Church from the early middle ages to the early modern period and accusations of sorcery and collusion with the devil, to Kant's program of "making room for faith," to the denunciation of "soulless materialism" in the nineteenth century, and of "reductionism" in the twentieth century, right down to demands for equal time for Creationism in the school curriculum and "vaccine hesitancy," science has faced mistrust. The dismal history of "scientific racism" and "scientific sexism," the exploitation of poor people and helpless animals for experiments, and the invention of demonic weaponry have put science "in the crosshairs." As Wolfgang Wieland, from whom I borrow my title, comments:

> Whoever is studying the history of science of modern times in the [crosshairs] of the enlightenment, will realize soon that science has always been in need of being illuminated about its own chances, risks, and side effects. The project of

enlightenment through science had to be complemented by the project of an enlightenment about science right from its beginning.[1]

To the accusations that the acquisition of scientific knowledge is distracting, depressing, heartless, and destructive of human values is joined a new one. Science claims the privileges of support, credence, and cultural respect, on the grounds that its methodology is uniquely suited to getting at the truth. But its claims are undercut, not just by philosophical analysis as to its knowledge claims[2] but by problems of validity, including those related to the reproducibility of results, experimental design, and statistical interpretation. Financial incentives and concerns about readership distort research and publication practices and popular journalism. It is like the old joke about resort dining—the food is terrible . . . and such small portions. Scientific knowledge is destructive—and it mostly isn't even true!

The focused attacks on scientism come in many distinct varieties. I first survey a set of five accusations against science which I designate as *Eclipse, Disenchantment, Fragmentation, Ideology, and Noise*. I argue that the purported offenses of science—scientistic harms—in these categories are either illusory or mostly self-healing. Further, the corrective course to scientistic harm in these cases arises *within* science, though the impetus to correct may come from philosophy and other sources of cultural criticism.

By contrast, two further categories of scientistic harm, *Misappropriation* and *No Boundaries*, should constitute the heart of the critique of scientism. They are not easily correctible from within because they are essentially moral criticisms and because they concern science as a well-rooted current practice rather than science as a set of changing representations. Although its practice involves many forms of normative regulation, and although its data can feed into moral evaluation, science is descriptive and explanatory; the values on which moral critique depends are wholly extraneous to it. Further, because *Misappropriation* and *No Boundaries* are in the self-interest of practitioners of science, their correction requires sacrifices.

1. SCIENTISM: SOME ACCUSATIONS

Eclipse. Rudolf Carnap is reported to have said that there is no [sensible] question that is not answerable by science, directly targeting the history of philosophy as a collection of "meaningless" statements that, at best, like poetry, expressed an "attitude toward life."[3] The contemporary *Eclipse* complaint against scientism is that scientific attitudes and practices, and the accompanying representations of the world, arrogate the whole field of human knowledge to themselves.

Carnap's accusations remain pertinent today. Until the late nineteenth century, natural and social science, and certain areas that we now demarcate as philosophy, notably philosophy of mind, philosophy of science, theory of agency, and political philosophy, were intertwined. As they began to come apart, the "linguistic turn" of analytic philosophy of the mid-twentieth century differentiated it sharply from empirical inquiry. Experimental and theoretical physics, linguistics, neuroscience, social studies of science, and experimental ethics claim to provide such knowledge of the traditional topics as is to be had.[4] Philosophers trained in analytic philosophy who attempt to theorize space and time, speech acts, perception, memory, free will, and moral judgment resentfully face claims that their efforts are works of the imagination and their disagreements no more substantial than the verbal play of medieval theological disputation.[5]

Disenchantment. Related to *Eclipse* is the accusation that science as an institution demeans non-scientific beliefs and practices—"other ways of knowing"—not only philosophical analysis and religious inspiration but also "intuition," feeling, aesthetic discrimination, and respect for the powers of nature. The "disenchantment of the world picture" of the neo-atomists of the seventeenth century and the "unweaving" of the rainbow by Newton were just the start.[6] Now the passions and ambitions of our friends, not to mention those of the great literary characters, are said to be explicable by hormones and neurotransmitters; love of children is explained by selection pressures operating early in the mammalian phylum; the experience of beauty is explained by cortical stimulation and associative mechanisms.[7] Thanks to progress in linguistic analysis and artificial intelligence, machines can compose and write poetry indistinguishable from that of human beings. The musical or poetic "genius" is not open to heaven; their neurons and synapses are just running a program. The scientists tell us that our personalities are determined by nature and nurture in ways beyond our personal control. It is not the stars that rule our lives and certainly not ourselves; it is the socioeconomic status of our parents, birth order, genetics, neuroanatomy, and physiology. This creates uneasiness, and critics of scientism retaliate with accusations of oversimplification.

Another version of Disenchantment is *Fragmentation*. Science dissolves the self into an illusion created by the brain.[8] We are portrayed as complicated robots who carry out the behavior the unconscious has computed as optimal under the circumstances. Our practices of responsibility-attribution and cultural valuing are, science tells us, works of the imagination, with no basis in reality.[9] This is unwelcome news to many moral philosophers, especially those concerned with "the agent," "the individual," the "first-person standpoint," "character," and other philosophical concepts.[10]

Ideology. The accusation here is that particular items of scientific research or communication overstep the boundaries of their author's expertise in order to push an obnoxious sociopolitical agenda. An example from recent history is the furor over Richard Dawkins' *Selfish Gene* of 1976. Dawkins' argument that human beings are machines built by genes for their own replication aroused horror in some philosophers and journalists and was interpreted as an argument excusing and so endorsing ruthlessly egoistic behavior as natural and unavoidable.[11] In the 1970s and 1980s, Robert Trivers' theory of differential parental investment by males and females was seen as suggesting that the role of males is to play, preen, fight, seduce, and force, and the role of females is to deal with it and its consequences, including, according to Randy and Nancy Thornhill, the human male propensity to rape.[12] A later furor arose over Richard J. Herrenstein's and Charles Murray's *Bell Curve* of 1993,[13] which claimed to have demonstrated the futility of certain interventions such as early childhood education (except as temporary relief from the dreadful home environment) and affirmative action, since the lower intelligence of African Americans was allegedly the principal cause of their low earnings, low status, and attendant criminality.

Noise. Science is "noisy" in the sense of Daniel Kahneman.[14] The *Journal of Irreproducible Results* was once an academic humor item, but it is becoming ever clearer how many results, especially in the social and behavioral sciences, are irreproducible.[15] Surely, you might think, the jury is in on the question of heart disease and cholesterol levels? Not so: The experts do not agree. Look at the medical literature on hormone therapy, Alzheimer's Disease, or fish oil supplements. Researchers attack one another's methodology and inferences. Controversy and retraction are rarely a matter of fraud and fudging; the control of variables, especially in the observational studies that necessarily dominate in medicine and poor experimental design are responsible for most problems.

The experts can get it badly wrong. In late March of the 2020 pandemic, the World Health Organization declared in a tweet "FACT: #COVID-19 is NOT airborne" and unleashed a mania for surface disinfection. Remembering perhaps the anthrax-in-the-mail attacks of 2001, the health correspondent of *NBC News* recommended a cumbersome set of procedures for dealing with the incoming mail such as wearing gloves while opening it and spraying or wiping it.[16] The WHO "FACT" was "updated" by a new fact in July stating the opposite. With scientific FACTS like these, everything is true![17] Conversely, because science is noisy, scientists are cautious when there is no immediate emergency and try to keep an open mind. For decades, researchers hedged on whether climate change was man-made, not because they didn't, among themselves, think so, but because standards of proof were high and the political and economic costs of being wrong would have been enormous.

2. RESPONDING TO CRITICISM:
SCIENCE SELF-CORRECTS

Where *Eclipse* is concerned, the conceptual analysis of academic philosophy—analytic metaphysics—is caught in a hard place. It deals with vernacular concepts, which different people understand differently, rather than the more precisely defined items of scientific jargon. Yet it is not interested in the untutored and varying judgments of the "man in the street" as elicited by surveys, but in the "armchair" intuitions of professional philosophers, refined by decades of mutual criticism as to actual entailment relations. Where the traditional topics of space, time, perception, memory, the aesthetic attitude, and free will are concerned, we do not know what its future will be. Analytic philosophy may fade into extinction, or it may become a form of popular science translating scientific theory into more intelligible forms; or it may persist as a discipline, akin to mathematics in having something like a proof structure, though without producing proofs regarded as so persuasive as not to be worth attacking.

Where *Disenchantment and Fragmentation* are concerned, the thoroughgoing replacement of common sense, traditional assumptions about the natural world and human relations, the implied causal accounts of human behavior furnished by myth, literature, and drama—and all the hopes, fears, and practices based on them—is unlikely. At the same time, as one of Shaw's characters says "You have learnt something. That always feels at first as if you had lost something." But if you have in fact learned it (i.e., if it is true), not only is your understanding enhanced but also your power to achieve what you want, given the actual pathways to fulfillment. Reality is what makes our plans succeed and fail.[18] To offer a simple analogy, it is disenchanting to learn that cherry tarts and sugary frappuccinos are not a healthy breakfast, but now, having lost that delicious illusion, you can take steps to reduce your risk of type 2 diabetes if that is a desired outcome.

Some alleged losses are compensated for by the re-enchantment offered by science, including the discoveries relating to the sensory world-constructing power of the brain, the explanation of dreaming and hallucination, the complexity of the microbiome, preadaptation and epigenetic mechanisms in evolution, hypotheses and discoveries relating to the early universe, and the power of cellular algorithms. When patiently explained by skilled interpreters, these discoveries do not produce feelings of alienation but of connectedness.

Where *Ideology* is concerned, science is self-correcting: that is its job. Dawkins had always insisted that highly transmissible human ideation—"memes"—frequently overruled drives for human survival and reproduction. On the biological level, it is now well understood that genes exist in contexts of many levels of cooperation and competition that determine their frequency in a population. The simplisms of parental investment theory—with coy monogamous females and aggressive polyamorous males capable of fathering "in principle" tens of thousands of offspring—have been corrected by the study of female choice and female infidelity, as well as logic.[19] And in the case of the *Bell Curve*, R. L. Leuwontin, Steven Rose, and Leon Kamin had already exposed some of the problems with measuring intelligence and making inferences from measurements, and Kamin and some eighty other authors weighed in with critique—and some support.[20] The American Psychological Association issued a statement distinguishing between the credible portions of the book and the noncredible.[21]

Scientists improve not only their beliefs about the world but their beliefs about how to improve their beliefs about the world, as Moti Mizrahi has pointed out.[22] The phenomenon of "long error" reflected in the scientific racism and sexism of previous centuries cannot occur in contemporary science in the absence of repression. Nevertheless, a spur to these corrections came from the cultural acquisition of moral perspectives and from social changes in the makeup of the scientific community that ultimately derived from the normative branches of philosophy.[23]

Noise is partly due to scientific incompetence or carelessness, partly to ambition and deception, but to a great extent it is simply due to the complexity of nature and the difficulty—now that the relatively low hanging fruit (difficult enough in its time to establish) has been picked off—of making new discoveries. Science aims to eliminate its own noise. There is an obvious need for vigilance in scientific publication and communication. But sorting the wheat from the chaff requires ever more powerful studies, on ever more elusive entities and interactions.

3. WHERE SCIENCE CAN'T SELF-CORRECT: A HARD LOOK AT THE SCIENTIFIC ESTABLISHMENT

Having dismissed this array of familiar criticisms of scientism as largely misguided, I turn to the most serious problems of contemporary scientific culture. As Wieland points out, "Although [individual scientists] are now free of the church authorities and censors who hounded them in the past," their choices are now controlled by "economic constraints and political dependencies." The interests of government and industry largely determine what is studied and even to some extent what is discovered, or reported as discovered, and what discoveries are applied and used.[24] These dependencies—the facilitation and discouragement they provide—may prove even more constraining on scientific autonomy than the old constraints of church and censor did, and they do not serve dependably human interests. Anti-scientism that arises from the suspicion or spectacle of what I will term *Misappropriation* and *No Boundaries* has a more forceful case to make than anti-scientism arising from the sources discussed earlier.

Misappropriation. The most egregious form of scientism involves simply spending too much money on science, thereby depriving other worthwhile endeavors of support and sustenance, as well as making the world a more dangerous and unpleasant place. Too much money in the developed world is directed to high-level scientific education and to the support of research and development, especially in warfare-related pursuits.

A scientific career can provide significant gratification and contribute to the economy even without producing results anymore "useful" than those of analytic metaphysics or French Studies. There is beauty and mystery in nature. Scientific problems can engage one's full concentration; they spark intuition and at the same time, like detective work, require patient clue-following and mechanism-unraveling. Scientists earn a living, purchase materials and journals, and pay taxes. But all endeavors in specialized subjects in the arts and humanities are rewarding in the same way, all provide employment, and all belong to the economy.

The issue is one of wildly unbalanced allocation, including allocations within different scientific fields. In the United States, most annual spending on science is furnished by industry (250 billion), followed by government (125 billion). In 2019, the US Department of Defense was allocated 55 billion for research and development; Health and Human Services, 38 billion. Much military research spending is hidden in other departmental accounts.[25] As one ex-researcher on "death rays" commented, "The military has a whole lot of money sloshing around, and they will try lots of different things, and some of them are good and some of them are not so good."[26] Despite its comparatively enlightened public face, the Department of Energy (17 billion) is less concerned with innovative green technologies than with the development, maintenance, and testing of nuclear weapons. NASA's budget is 11 billion and that of the National Science Foundation is around 6 billion. By contrast, President Trump proposed to cut the research budget of the Environmental Protection Agency by 42 percent to 242 million and that of the Department of Education to 224 million. Even without these cuts, the miserliness of allocations for projects directly related to human welfare is staggering. And how is it possible that the National Institutes of Health (Motto: "Turning Discovery into Health") can spend 42 billion per year, while health outcomes in the United States are dramatically worse in terms of chronic disease, avoidable death, and life expectancy than in comparable OECD countries?[27]

One may well ask: do we need all this applied scientific research? Where industry is concerned, are our consumer products—our cosmetics and shampoos, pills and potions, TVs and refrigerators—perhaps "good enough"? Where government is concerned, is our collection of helicopters, bombs, missiles, and nuclear submarines perhaps good enough for national defense without new resources being poured into their improvement? The relentless search for the new and improved is exceedingly costly to the environment and feeds the arms race without enhancing security. Although Covid-19 pandemic provided a dramatic example of the power of well-funded scientific innovation to save lives, many basic human needs for housing, childcare, and healthcare that could be satisfied without new scientific discoveries are not met because they are allegedly too expensive (though "expensive" implies job producing and tax revenue producing). One might think for a society that can send probes to Jupiter and vehicles to Mars (18 billion and counting)[28] nothing is too expensive.

One need not begrudge NASA its budget. My back of the envelope calculation, subject to correction, is that NASA costs the average taxpayer only $76 per year. My point is that other forms of spending and subsidization that provide jobs in the caring and repairing professions that are of similarly low cost to the individual taxpayer are deemed unaffordable. Cleaning up toxic waste dumps, restoring marshlands and forests, and taking care of three-year-olds so their mothers can enjoy meaningful work in the company of other adults, (which science tells us is embedded in female nature)[29] are not seen as cool and fun, like exploring Mars.

In the universities, STEM subjects receive from the government alone about thirteen times as much funding as humanities.[30] Former President Trump proposed shutting down the National Endowment for the Humanities and the National Endowment of the Arts during his time in office; they have since been reprieved. The success rate of grant applications to the National Institute of Health, with, as noted, a total budget of 42 billion, is about 20–35 percent. The success rate of applications to the NEH with a total budget of about 152 million (.003 of the above figure), is about 16 percent. These figures obviously reflect perceptions of "worthwhileness" on the part of decision-makers, but it has to be asked whose perceptions count and what is their basis. To what extent are these perceptions of a need for more and more science holdovers from the cold-war panic of the last century and the results of lobbying and propaganda, rather than a thoughtful reflection on human interests, needs, and abilities?

For the last several decades, universities have deliberately shifted resources from the arts and humanities into STEM subjects. They effectively transfer the proceeds of the long teaching and student assessment hours of arts and humanities that make these courses rewarding to our students to the support of overheads and teaching release in the sciences. Subsidized by the humanities and by cost savings from the closure and contraction of programs, the return on investment in academic scientific research can appear more impressive than it really is.[31]

Is there a labor shortage of qualified scientific researchers such that extraordinary measures are needed to recruit and teach them? This is a myth believed even by those who should know better. The summit of 2003 convened to study this question found no evidence of any such absolute shortage; the only concern was that other countries were churning out more scientists.[32] On the question of whether having more science BAs and PhDs in the population would contribute to the well-being of humanity, or even the nation, the 2003 commission was evasive, having no real idea how to answer that question. According to Alan Greenspan, the former Federal Reserve Chairman, in his 2000 testimony to Congress, more training of scientists was needed for the industry to keep its wage costs low through an overabundance of qualified applicants for jobs.[33] It was in the more forgotten part of his speech that Greenspan went on to say,

> Many academics argue, I believe rightly, that significant exposure to a liberal education—music, literature, and the arts—broadens intellectual awareness, enhancing the ability to reach across disciplines to forge new ideas. Thus, while we must strengthen math and science education to address the requirements of the newer technologies we see on the horizon, we should not lose sight of the advantages of a liberal education.

Contrary to what is often assumed, attitudes toward the humanities among the US populace, while not overwhelmingly favorable compared to attitudes toward STEM subjects, are not overwhelmingly hostile: very few Americans consider the humanities a complete waste of time or objectionably elitist.[34]

No Boundaries. Scientific progress is measured not only by an increase in understanding but by an increase in control over nature and redirection of nature's course. "The contemplative ideal of scientific investigations for their own sake has been replaced in modern times," Wieland comments, "by the practical ideal of scientific research in the service of humanity."[35] The natural course of things once presented and still presents threats to human welfare: people went and still go hungry, die in infancy and early adulthood, and face natural disasters. But not all control and redirection add to human welfare. Medical technology, based on science, has lengthened the number of years human beings suffer from chronic diseases, condemning them to lonely, bedridden existences. End-of-life treatments that prove medical innovation and competence can sustain life, but not living. The illustrious medical journal, the *Lancet*, reports frequently on the "sea of suffering" in aging populations.[36]

The threats of technological dystopias are real: Do we really want the rich and powerful to live to 150 through life extension interventions controlling the world as they see fit? Do we want our great-grandchildren to be cyborgs? Growing artificial brains is cool and fun, as would be filling the sky with sunlight-reflecting plastic beads to combat global warming or walking on Mars. For too many scientists, the attitude is: "If they will fund it, I will do it!" Grant writers are skilled in proposing humanitarian applications, and evaluators are skilled in seeing through them. But if allocations are there, they must be spent. In the defense and chemical industries, scientists are rarely enlisted to immoral causes because of their own biases and convictions; more usually it is because of the need to earn a living, and the cognitive and emotional labor involved in trying to envision and reflect on potential consequences and their moral value. That is conceived as someone else's specialty and someone else's job, even while the hotbeds of those reflective and representational specialties, in the arts, literature, and the humanities, are drained of resources. With roughly half of the research funding coming from the defense agencies, individual scientists who benefit are reluctant to take a stance against the over-militarization of the world. In the life sciences, ethics committees operate within very broadly construed conceptions of useful knowledge and are reluctant to turn down the proposals of prominent researchers.[37]

The moral irresponsibility of scientists who ask themselves only "Would it be interesting and challenging to find this out/make this?" needs to be put in the crosshairs. Scientists may tell themselves that their role is that of a pure search for understanding, that there are no ethical implications as such to their discoveries, and that it is up to entrepreneurs, agencies, legislators, and voters to decide what to pursue and how to apply their findings. They are right about implication in its logical sense: "is" does not imply "ought" in the absence of context. But it assuredly does in context, that is, "The pilot is drunk; therefore, he should not fly the plane." There is arguably a moral duty for the investigator him or herself to envision, and to be limited in deciding what research to pursue, by the recognition of the worst possible applications of discoveries that can be foreseen from the present standpoint. The responsibility of putting the brakes on research should not be turned over entirely to others with a less accurate grasp of its potential. A refusal to cooperate with the death-dealing industries, no matter how intriguing and lucrative their research problems, is another indication of moral courage.

There are promising signs of a greater concern with the ethics of scientific practice. Individuals discuss over blogs the qualms that have led them to abandon weapons research. A recent editorial in *Nature* is resigned to the military funding of university research as inevitable. The author insists nevertheless that "the researchers involved carry a heavy responsibility. The work should align with a fundamental commitment to humane and life-saving applications—drones that can deliver medical supplies to war-torn areas, or robots that can clear minefields, for example."[38]

4. TOO MUCH . . . BUT ALSO, NOT ENOUGH! SCIENCE, NOT SCIENTISM

The original promise of science in the period of the seventeenth- and eighteenth-century scientific revolution was, on one hand, to release human beings from irrational fears and superstitious beliefs, such as the belief in witches, Hell, and portents, and, on the other, to enhance the ability to predict the future, and to furnish us with useful and desirable products, especially medicines, gold, and military technology. It has fulfilled its promise in all these cases—though not by transmutation of base metals. But natural and social scientists know an enormous amount not falling under these categories that could enhance human welfare were their knowledge to get through to the holders of power and to be found convincing by them.

Suppose as a thought experiment, we set aside considerations of the economy, political expediency, tradition, and the fascination with innovation and consumer choice—admittedly things that matter deeply to people—and focus on what science could do to make our lives healthier, more secure, and more pleasurable without infringing on important liberties. If scientific knowledge is indeed the best knowledge we have,[39] what would "following the science"—both natural and social—advise us to do? Probably the following:

1. Curtail exploration for new fossil fuels.
2. Focus theory and practice on climate mitigation and restoring nature.
3. Ban chemicals in plastics, pesticides, and herbicides that pose dangers to human health.
4. Stop producing junk food, reduce though not eliminate meat production and consumption.
5. Prescribe drugs and medical procedures by reference to absolute, not just relative, risk. Control the overuse of antibiotics that spur the development of resistance.
6. Ban the manufacture and sale of guns to private individuals on the grounds of what is known about the risks of homicide, suicide, and mental illness.
7. Modify the customs and legal arrangements of marriage and divorce to be accepting of the emotional and cognitive features of human thought and behavior and the different stages and needs of the human life cycle.

8. Draw on the sociological and economic understanding of poverty to move beyond the simplistic personal responsibility vs. parasitism debate and implement effective policies.
9. Overhaul the criminal justice system in light of what is known about the inefficaciousness of most punishment and the efficacity of some but not all methods of rehabilitation.
10. Employ what is known about blackmailers and negotiation strategies to deal with rogue nations and intransigent dictators with less terror and bloodshed.

The moral desiderata reflected here arise outside of science; they are not justified by science. But the problem is not that these desiderata are deeply controversial unlike the long-held consensus results of science. It is that despite the uncontroversial nature of the values of life, health, security, and pleasure, the gap between scientific knowledge and political practice is a large one. This raises questions of knowledge transfer.

This important concept is largely understood in connection with inter-laboratory exchanges and with industrial, not political, applications. But the latter are supremely important. Governments at all levels from local to international seek scientific advice, but when this advice is not politically expedient it can be ignored as corresponding only to scientific opinion or mere theory. Contrarians can always be scratched up to sign off on the minority opinion, which, thanks to the intrinsically nondogmatic and self-critical nature of scientific inquiry, can hold up policy decisions indefinitely. Scientific open-mindedness and caution thus undermine its epistemic claim to generate action-guiding knowledge.

The obstacles to effective knowledge transfer include, in addition to *Noise*, faulty information (including cherry-picking from noisy research areas), generational differences, resistance to change, power struggles, and above all, "the capabilities of the receptor to interpret and absorb knowledge."[40] Further, there will always be a core of resistance to the enhancements of the general welfare that science could give us because this would involve a sacrifice of some of the values of choice, expediency, and wealth-production mentioned above. Moral change is difficult because it is, "advantage reducing": it involves sacrifices from the powerful on behalf of the powerless which do not serve their short-term interests.

To conclude, science, though not scientism, can put to rest the most delusory and destructive ideas of human beings about one another and their world in a way that no other field of inquiry can accomplish on its own. A scientific education at the elementary and high school levels that would instill the basic principles of empiricism and critical thinking (rather than the details of the periodic table and the insides of the frog) is a desperate need in a scientifically confused but at the same time overly acquiescent society and is more important to social progress than the training of more research scientists. We nonscientists need the hard-won contact with reality—or at any rate, the empirical adequacy—that science aims at. The normative aims of the satisfaction of real and legitimate needs and the prevention of suffering can only be accomplished through a combination of philosophical reflectiveness and critique with scientific knowledge.

NOTES

1. Wolfgang Wieland, 'Wissenschaft im Fadenkreuz der Aufklärung.'Zur Tragweite des hypothetischen Denkens [Science in the crosshairs of enlightenment. On the significance of hypothetical thinking].' *Acta Hist Leopoldina* 57 (2011):99–130. Abstract retrieved from https://pubmed.ncbi.nlm.nih.gov/22397137/.

2. See Bas Van Fraassen's account of constructive empiricism in *The Scientific Image*, Oxford: Oxford University Press, 1980.

3. Rudolf Carnap, 'The Elimination of Metaphysics through Logical Analysis of Language,' tr. Arthur Pap. In A.J. Ayer, ed., *Logical Positivism,* Free Press: Glencoe IL, 60–81.

4. For these accusations, see Tom Sorell, *Scientism: Philosophy and the Infatuation with Science*. London: Routledge, 1991.

5. See Richard Rorty, 'Philosophy as a Kind of Writing: An Essay on Derrida.' *New Literary History*, 10: 1(1978) pp. 141–160.

6. Max Horkheimer, and Theodor Adorno, *'The Concept of Enlightenment.'* In *The Dialectic of Enlightenment*, Stanford: Stanford University Press, 1964, 1–34; See Lorraine Daston, 'History of Science in an Elegiac Mode: EA Burtt's Metaphysical Foundations of Modern Physical Science Revisited,' *Isis* 82.3 (1991): 522–531.

7. See the critique by John Hyman, 'Art and Neuroscience.' In *Beyond Mimesis and Convention.*, ed. Roman Frigg and Matthew Hunter, Dordrecht: Springer, 2010, 245–261.

8. Bruce Hood, *The Self Illusion*: *How the Social Brain Creates Identity*. Oxford: Oxford University Press, 2012.

9. D.M. Wegner, *The Illusion of Conscious Will*. Cambridge MA: MIT Press, 2003; Max Velmans, 'Is Human Information Processing Conscious? *Behavioral and Brain Sciences* 4.4 (1991): 651–669.

10. See John Doris' psychology-based attack on the philosophical notion of moral character in *Lack of Character: Personality and Moral Behavior*, Cambridge: Cambridge University Press, 2005.

11. Mary Midgely, 'Gene-juggling,' *Philosophy* 54.210 (1979): 439–458. Cf. Richard Dawkins's reply 'In Defence of Selfish Genes.' *Philosophy* 56.218 (1981): 556–573.

12. Robert Trivers. 'Parental Investment and Sexual Selection.' In *Sexual Selection in the Descent of Man,* ed. Bernard G. Campbell, New York: Aldine de

Gruyter, 1972: 136–179. Randy Thornhill and Nancy Wilmsen Thornhill. 'Human Rape: An Evolutionary Analysis.' *Ethology and Sociobiology* 4.3 (1983): 137–173.

13. Charles Murray and Richard J. Herrnstein, *The Bell Curve: Intelligence and Class Structure in American life*. NY: Free Press, 1994.

14. Daniel Kahneman, et. al. 'Noise.' *Harvard Business Review* October 2016: 38–46.

15. Stuart Ritchie, *Science Fictions: How Fraud, Bias, Negligence, and Hype Undermine the Search for Truth*, New York: Macmillan, 2020.

16. Madelyn Fernstrom, on "best practices" "As states open up, should you sanitize your mail and packages? Here's your COVID-19 home delivery guide." Retrieved from https://www.nbcnews.com/know-your-value/feature/states-open-should-you -sanitize-your-mail-packages-here-s-ncna1217371.It is doubtful that such essentially made-up-for-the-article procedures saved a single person from infection.

17. The WHO's erroneous reasoning—replicated in many textbooks—was revealed by a University of Virginia aerosol researcher, Linsey Marr. See Megan Molteni, 'The 60 Year Old Scientific Screw up than Helped Covid Kill,' *Wired* 05.13.2021. Retrieved from https://www.wired.com/story/the-teeny-tiny-scientific -screwup-that-helped-covid-kill//

18. According to White's condensation of Frank Ramsey's formulation, "A belief 's truth condition is that [state of the world] which guarantees the fulfillment of any desire by the action which that belief and desire would combine to cause." See J.T. Whyte, 'Success semantics,' *Analysis*, 50 (1990), 149–157.

19. The number of offspring engendered by the average male is the same as the number engendered by the average female in conditions of an equal sex ratio. "Reproductive skew"—the variance in the number of offspring among males in the environment of early adaptation is held to have been small relative to other primates and relative to the opportunities afforded by early civilizations that enabled harem building. See Laura Betzig. "Means, variances, and ranges in reproductive success: comparative evidence." *Evolution and Human Behavior* 33.4 (2012): 309–317.

20. See Leon Kamin, 'Lies, Damned Lies, and Statistics.' In Russell Jacoby, and Naomi Glauberman, eds. *The Bell Curve Debate,* New York: Times Books (1995): 86–105. Cf. Richard C., Lewontin, Steven Rose, and Leon J. Kamin. *Not in our Genes: Biology, Ideology, and Human Nature*. New York, NY: Pantheon Press 1993.

21. Ulric Neisser, et al. *Intelligence: Knowns and Unknowns*, Washington, DC: American Psychological Association, 1995. Retrieved from http://citeseerx.ist.psu .edu/viewdoc/download?doi=10.1.1.322.5525&rep=rep1&type=pdf).

22. Mizrahi points to the need to include "methodological knowledge," in addi-tion to empirical, theoretical, and practical knowledge, in a full account of scientific progress. "What is Scientific Progress? Lessons from Scientific Practice," *Journal for General Philosophy of Science*, 44 (2013) 375–90, p. 380.

23. See note xix. These corrections were arguably an effect of the long-delayed entrance of female researchers into the academy.

24. Wieland, "Wissenschaft im Fadenkreuz" (Abstract).

25. "Late-stage development, testing, and evaluation primarily within the Department of Defense" are no longer counted as R & D. American Association for the Advancement of Science, 'Trends in Federal R & D FY 1976–2020,' Retrieved from https://www.aaas.org/sites/default/files/2020-10/DefNon.png.

26. Cheryl Rofer, quoted in Julian Borger, 'Microwave Weapons that could cause Havana Syndrome Exist, Experts Say,' *Guardian*, 2 June 2021. Retrieved from https://www.theguardian.com/science/2021/jun/02/microwave-weapons-havana-syndrome-experts

27. Commonwealth Fund, 'U.S. Health Care from a Global Perspective, 2019: Higher Spending, Worse Outcomes, 'Issue Briefs, January 2020. Retrieved from https://www.commonwealthfund.org/publications/issue-briefs/2020/jan/us-health-care-global-perspective-2019.

28. World Economic Forum, 'Chart. This is how much each of NASA's Mars missions have cost.' https://www.weforum.org/agenda/2021/02/mars-nasa-space-exploration-cost-perseverance-viking-curiosity/. Defending annual NASA expenditures Jeff Kerns, in a blog not I think intended to be subversive, points out that space research also "invented the battery technology that allows for cordless vacuums and other tools, baby food, freeze-dried technology, artificial limbs, powdered lubricants, and hook and loop fasteners (Velcro). That list doesn't include the inventions indirectly generated around NASA, such as the Super Soaker." Jeff Kerns, 'Is NASA a waste of money?' [Blog Post] (2016) Retrieved from https://www.machinedesign.com/community/editorial-comment/article/21832319/is-nasa-a-waste-of-money.

29. For a recent update, see Marlize Lombard and Katherine Kyriacou, 'Hunter-Gatherer Women.' *Oxford Research Encyclopedia of Anthropology,* ed. Mark Aldenderfer, Retrieved from https://oxfordre.com/anthropology/view/10.1093/acrefore/9780190854584.001.0001/acrefore-9780190854584-e-105.

30. Jeffrey Brown, Julia Hayes, Simon Rhodes, Cathleen Webb, 'Federal Funding Sources: Where the Money Really Is...' Retrieved from https://www.ccas.net/files/2015%20Annual%20Meeting%20Washington%20DC/Presentations/Federal%20Funding%20Sources.pdf . According to data "released by the American Academy of Arts & Sciences this summer, the government pays for well over 50 percent of the scientific research done in universities, and close to 75 percent in some disciplines. Meanwhile, the humanities are fronting all but 20 percent of their own costs." Nora Caplan-Bricker, 'New Evidence: There is no Science Education Crisis,' *New Republic*, September 5, 2013. Retrieved from https://newrepublic.com/article/114608/stem-funding-dwarfs-humanities-only-one-crisis.

31. "The senior managers at ASU, like everywhere else, keep the university solvent by taking the SASH surpluses on the right and using them to fill the STEM budget holes on the left." Christopher Newfield, 'The Humanities as Service Departments: Facing the Budget Logic,' *MLA Profession*, December 2015 Retrieved from https://profession.mla.org/the-humanities-as-service-departments-facing-the-budget-logic/ "The conclusion here," Newfield adds, "is that STEM profits depend on SASH subsidies. Ignoring subsidies artificially elevates ROI."

32. William P. Butz, Gabrielle A. Bloom, Mihal E. Gross, Terrence K. Kelly, Aaron Kofner, Helga E. Rippen, 'Pan-Organizational Summit on the US Science and Engineering Workforce: Meeting Summary, 2003.' Retrieved from https://www.ncbi.nlm.nih.gov/books/NBK36380/#a2000882addd00166. More recently, see Beryl Lieff Benderly, 'What Scientist Shortage?' *Columbia Journalism Review.* Jan-Feb. 2012. Retrieved from https://archives.cjr.org/reports/what_scientist_shortage.

33. Alan Greenspan, 'The Economic Importance of Improving Math-Science Education,' Speech before the Committee on Education and the Workforce, U.S. House of Representatives September 21, 2000. Retrieved from https://www.federal-reserve.gov/boarddocs/testimony/2000/20000921.html

34. See the survey of the American Academy of Arts and Sciences, 'How Americans View the Humanities,' Retrieved from https://www.amacad.org/publication/humanities-american-life/section/4

35. Wieland, 'Wissenschaft im Fadenkreuz' [Abstract]. While nobody has any idea what the human welfare spin-offs of the Large Hadron Collider will be, the question gets asked frequently in the media. "We are just curious to learn about nature," however authentic a response, is not usually taken as justifying the project.

36. Richard Horton, 'A sea of suffering' *The Lancet,* 391: 10129 (2019): 1465, and Editorial, 'Thinking chronic pain,' *The Lancet,* 397: 10289 (2021)

37. Julian Savulescu, Iain Chalmers, and Jennifer Blunt, 'Are research ethics committees behaving unethically? Some suggestions for improving performance and accountability.' *British Medical Journal,* 313:7069 (1996): 1390–1393.

38. 'Military work threatens science and security,' Editorial, *Nature* 556: (17 April 2018) 273. Retrieved from doi: https://doi.org/10.1038/d41586-018-04588-1

39. See for a defence of this claim on statistical as well as conceptual grounds, Moti Mizrahi, 'What's so Bad about Scientism?' Social Epistemology (2017) 31.4: 351–67.

40. Wikipedia Art. 'Knowledge Transfer.' https://en.wikipedia.org/wiki/Knowledge_transfer

REFERENCES

American Academy of Arts and Sciences. *How Americans View the Humanities.* https://www.amacad.org/publication/humanities-american-life/section/4.

American Association for the Advancement of Science. *Trends in Federal R & D' FY 1976–2020.* https://www.aaas.org/sites/default/files/2020-10/DefNon.png.

Benderly, Beryl Lieff. "What Scientist Shortage?" *Columbia Journalism Review* (Jan–Feb. 2012). https://archives.cjr.org/reports/what_scientist_shortage.php.

Betzig, Laura. "Means, Variances, and Ranges in Reproductive Success: Comparative Evidence." *Evolution and Human Behavior* 33.4 (2012): 309–317.

Borger, Julian. "Microwave Weapons That Could Cause Havana Syndrome Exist, Experts Say." *Guardian,* 2 June 2021. https://www.theguardian.com/science/2021/jun/02/microwave-weapons-havana-syndrome-experts.

Brown, Jeffrey, Julia Hayes, Simon Rhodes, and Cathleen Webb. "Federal Funding Sources: Where the Money Really Is...." *Conference Presentation.* https://www.ccas.net/files/2015%20Annual%20Meeting%20Washington%20DC/Presentations/Federal%20Funding%20Sources.pdf.

Butz, William P., Gabrielle A. Bloom, Mihal E. Gross, Terrence K. Kelly, Aaron Kofner, and Helga E. Rippen. *Pan-Organizational Summit on the US Science and Engineering Workforce: Meeting Summary, 2003.* https://www.ncbi.nlm.nih.gov/books/NBK36380/#a2000882addd00166.

Caplan-Bricker, Nora. "New Evidence: There is No Science Education Crisis." *New Republic*, 5 September 2013. https://newrepublic.com/article/114608/stem-funding-dwarfs-humanities-only-one-crisis.

Carnap, Rudolf. "The Elimination of Metaphysics Through Logical Analysis of Language." Translated by Arthur Pap. In *Logical Positivism*, edited by A. J. Ayer, 60–81. Free Press: Glencoe, IL.

Commonwealth Fund. "U.S. Health Care from a Global Perspective, 2019: Higher Spending, Worse Outcomes." *Issue Briefs*, January 2020. https://www.commonwealthfund.org/publications/issue-briefs/2020/jan/us-health-care-global-perspective-2019.

Daston, Lorraine. "History of Science in an Elegiac Mode: EA Burtt's Metaphysical Foundations of Modern Physical Science Revisited." *Isis* 82.3 (1991): 522–531.

Dawkins, Richard. "In Defence of Selfish Genes." *Philosophy* 56.218 (1981): 556–573.

Doris, John. *Lack of Character: Personality and Moral Behavior.* Cambridge: Cambridge University Press, 2005.

Editorial. "Thinking Chronic Pain." *The Lancet* 397.10289 (2021).

Editorial Board. "Military Work Threatens Science and Security." *Nature* 556.7701 (17 April 2018): 273. DOI: 10.1038/d41586-018-04588-1.

Editorial Board. "Thinking Chronic Pain." *The Lancet* 397 (2021): 10289.

Fernstrom, Madelyn. *As States Open Up, Should You Sanitize Your Mail **and** Packages? Here's Your COVID-19 Home Delivery Guide.* https://www.nbcnews.com/know-your-value/feature/states-open-should-you-sanitize-your-mail-packages-here-s-ncna1217371.

Greenspan, Alan. "The Economic Importance of Improving Math-Science Education." Speech Before the Committee on Education and the Workforce, U.S. House of Representatives September 21, 2000. https://www.federalreserve.gov/boarddocs/testimony/2000/20000921.htm.

Hood, Bruce. *The Self Illusion: How the Social Brain Creates Identity.* Oxford: Oxford University Press, 2012.

Horkheimer, Max, and Theodor Adorno. "The Concept of Enlightenment.' In *The Dialectic of Enlightenment*, 1–34. Stanford: Stanford University Press, 1964.

Horton, Richard. "A Sea of Suffering." *The Lancet* 391.10129 (2019): 1465.

Hyman, John. "Art and Neuroscience." In *Beyond Mimesis and Convention*, edited by Roman Frigg and Matthew Hunter, 245–261. Dordrecht: Springer, 2010.

Kahneman, Daniel, Andrew M. Rosenfield, Linnea Gandhi, and Tom Blaser. "Noise: How to Overcome the High, Hidden Cost of Inconsistent Decision Making." *Harvard Business Review* (October 2016): 38–46.

Kamin, Leon. "Lies, Damned Lies, and Statistics." In *The Bell Curve Debate*, edited by Russell Jacoby and Naomi Glauberman, 86–105. New York: Times Books, 1995.

Kerns, Jeff. *Is NASA a Waste of Money?* 2016 [Blog Post]. https://www.machinedesign.com/community/editorial-comment/article/21832319/is-nasa-a-waste-of-money.

Lewontin, Richard C., Steven Rose, and Leon J. Kamin. *Not in Our Genes: Biology, Ideology, and Human Nature*. New York, NY: Pantheon Press, 1993.

Lombard, Marlize, and Katherine Kyriacou. "Hunter-Gatherer Women." In *Oxford Research Encyclopedia of Anthropology*, edited by Mark Aldenderfer. https://oxfordre.com/anthropology/view/10.1093/acrefore/9780190854584.001.0001/acrefore-9780190854584-e-105.

Midgely, Mary. "Gene-Juggling." *Philosophy* 54.210 (1979): 439–458.

Mizrahi, Moti. "What is Scientific Progress? Lessons From Scientific Practice." *Journal for General Philosophy of Science* 44 (2013): 375–390, 380.

Mizrahi, Moti. "What's So Bad About Scientism." *Social Epistemology* 31.4 (2017): 351–367.

Molteni, Megan. "The 60 Year Old Scientific Screw Up Than Helped Covid Kill." *Wired*, 13 May 2021. https://www.wired.com/story/the-teeny-tiny-scientific-screwup-that-helped-covid-kill/.

Murray, Charles, and Richard J. Herrnstein. *The Bell Curve: Intelligence and Class Structure in American life*. New York: Free Press, 1994.

Neisser, Ulric, Gwyneth Boodoo, Thomas J. Bouchard, Jr., A. Wade Boykin, Nathan Brody, Stephen J. Ceci, Diane F. Halpern, J. C. Loehlin, R. Perloff, R. J. Sternberg, and S. Urbina. *Intelligence: Knowns and Unknowns*. Washington, DC: American Psychological Association, 1995. http://citeseerx.ist.psu.edu/viewdoc/download?doi=10.1.1.322.5525&rep=rep1&type=pdf.

Newfield, Christopher. "The Humanities as Service Departments: Facing the Budget Logic." *MLA Profession*, December 2015. https://profession.mla.org/the-humanities-as-service-departments-facing-the-budget-logic/.

Ritchie, Stuart. *Science Fictions: How Fraud, Bias, Negligence, and Hype Undermine the Search for Truth*. New York: Macmillan, 2020.

Rorty, Richard. "Philosophy as a Kind of Writing: An Essay on Derrida." *New Literary History* 10.1 (1978): 141–160.

Savulescu, Julian, Iain Chalmers, and Jennifer Blunt. "Are Research Ethics Committees Behaving Unethically? Some Suggestions for Improving Performance and Accountability." *British Medical Journal* 313.7069 (1996): 1390–1393.

Sorell, Tom. *Scientism: Philosophy and the Infatuation With Science*. London: Routledge, 1991.

Thornhill, Randy, and Nancy Wilmsen Thornhill. "Human Rape: An Evolutionary Analysis." *Ethology and Sociobiology* 4.3 (1983): 137–173.

Trivers, Robert. "Parental Investment and Sexual Selection." In *Sexual Selection in the Descent of Man*, edited by Bernard G. Campbell, 136–179. New York: Aldine de Gruyter, 1972.

Van Fraassen, Bas. *The Scientific Image*. Oxford: Oxford University Press, 1980.

Wegner, D. M. *The Illusion of Conscious Will.* Cambridge, MA: MIT Press, 2003.

Velmans, Max. "Is Human Information Processing Conscious?" *Behavioral and Brain Sciences* 4.4 (1991): 651–669.

Whyte, J. T. "Success Semantics." *Analysis* 50 (1990): 149–157.

Wieland, Wolfgang. "Wissenschaft im Fadenkreuz der Aufklärung. Zur Tragweite des hypothetischen Denkens [Science in the Crosshairs of Enlightenment. On the Significance of Hypothetical Thinking]." *Acta Hist Leopoldina* 57 (2011): 99–130. Abstract https://pubmed.ncbi.nlm.nih.gov/22397137/.

Wikipedia, Art. *Knowledge Transfer.* https://en.wikipedia.org/wiki/Knowledge_transfer.

World Economic Forum. *Chart. This Is How Much Each of NASA's Mars Missions Have Cost.* https://www.weforum.org/agenda/2021/02/mars-nasa-space-exploration-cost-perseverance-viking-curiosity/.

Chapter 8

Between Electrical Light Switches and Panpsychism

Scientism and the Responsibilities of the Humanities in the Twenty-First Century

Ann-Sophie Barwich

Can we use light switches and, at the same time, believe in myths? This question resonates with ongoing disputes about the authority of science versus non-scientific ways of thinking. Recently, concerns regarding an overreach of scientific authority in human culture renewed momentum to pseudoscientific ideas originating in anti-science sentiments. This chapter sets out to rethink the currently prevailing image of science in light of its role as a facilitator of cognitive evolution to counter these potentially harmful conjectures. My argument unfolds in three steps. It first touches on the specific role of science in human cultural development. Drawing on ideas by Bultmann of the Marburg school, I suggest distinguishing between cosmological and existentialist questions in our engagement with the world. This distinction forms the backdrop against which we can understand the current popularization and rise of speculative metaphysics. Second, I examine a specific example: the recent revival of panpsychism as an example of an increasing conflation of existentialist with cosmological questions. Of particular concern in this context is that current support for panpsychism is fueled by a science-skepticism that draws on arguments from the philosophy of science. In a third step, this chapter warns that these arguments used to back science-skepticism build on a misleading image of science, which portrays science as an unchanging cognitive practice with varying knowledge outputs. However, I contend that science is not principally in the business of accumulating facts; rather, it is a cognitive activity in the pursuit of more sophisticated questions for understanding the world. Such shift in the image of science results in a change of its explanatory target by focusing on cognitive participation in its processes

139

instead of centering analysis on its products like knowledge and technologies. Linking to recent theories of cultural evolution in cognitive science, analysis of science is best framed via the evolutionary development of *cognitive gadgets* or mental mechanisms through cultural transmission. Such a revised image of science then offers new avenues of collaboration with the humanities in the twenty-first century.

1. INTRODUCTION: LIGHT SWITCHES AND MIRACLES

Can you use a light switch, a vaccine, or a computer at the same time as holding metaphysical beliefs in conflict with a scientific causal-mechanistic worldview? In 1941, Rudolf Karl Bultmann, a theology professor at Marburg, gave a negative verdict. Revisiting the status of the mythology in the New Testament, he concluded:

> We cannot use electric lights and radios and, in the event of illness, avail ourselves of modern medical and clinical means and at the same time believe in the spirit and wonder of the New Testament. (Bultmann 1989, 4)

Bultmann did not eschew science. Acceptance of Christian metaphysics—with its existence of heaven, earth, and hell—seemed incompatible with modern cosmology. Influenced by Heidegger and Nietzsche, Bultmann saw a need to rethink the existential role of belief and theology in the twentieth century. Responses to Bultmann were divided. Critical engagement with his essay "New Testament and Mythology: The Problem of Demythologizing the New Testament Proclamation" spanned several decades throughout the 1940s to the 1980s, for example, in the series *Kerygma and Mythos*. Vociferous critics claimed that they had found "signs of senility" (Hans Joachim Iwand quoted in Labron 2011). Others still wonder: "is [Bultmann] a good influence or a bad influence, is he a defender of orthodoxy or is he a heretic?" (Labron 2011, 7).

Bultmann's thinking is of broader interest than a theological dispute in its attempt to resolve the tension that characterizes the modern mindset as trapped between myths and method, miracles and light switches. It touches upon an ongoing social dilemma: the role of non-scientific ways of engagement with a world primarily governed by technological advances that are rooted in a scientific understanding of matter, causes, and effect.

Science as the central driver of modern society, emphasizing the importance of technological gadgets and the image of evidence-based reasoning, has attracted its fair share of critics. Scholarship in Science and Technology Studies and the humanities has an especially long history of cautioning about the lure of "Scientism": the lure that we place "too high a value on natural science in comparison with other branches of learning or culture" (Sorrell 1991; quoted in Burnett 2014) and the misconception that science "is the only source of real knowledge" (Hutchinson 2011; quoted in Burnett 2014). This view of "Scientism" cautions about the overreach of scientific authority in human society and should be distinguished from other philosophical understandings of "scientism," which refer to epistemological foundations of scientific knowledge, such as Ladyman (2008).[1,2] Critics of the view of Scientism as an overreach of authority have highlighted the importance of cultural knowledge residing outside scientifically oriented inquiry and noted on the historicity of science itself, including its social placement. Another point of concern involves the theory-ladenness of scientific observation, characterizing scientific understanding as inevitably perspectival in comparison with a culture-bound view of the world. Accordingly, ignoring the historical and cultural situatedness of science may resemble religion or pseudoscience more than the tenets of science itself (Burnett 2014; Pigliucci 2018). Meanwhile, valid criticism of Scientism can adopt an uncomfortably moralizing stance, with comments about the "arrogance and intellectual bullyism" of science sprinkled in (Hutchinson 2011; quoted in Burnett 2014). Moreover, this outlined dualism between scientific and non-scientific ways of engagement with the world occasionally gets hyped into polemical heights for clicks and readership, so much so that one may wonder how much understanding of science really is needed to form judgments about its character. In what follows, I thus want to engage with the danger of science illiteracy in this context to offer an alternative to this mounting opposition of scientific versus non-scientific thinking.

The rivalry of "the Two Cultures" seems to rest on a distorted and ahistorical image of science (and, in turn, the humanities). I want to offer an alternative picture by putting forth three claims: First, drawing on Bultmann's thinking, it is vital to distinguish between cosmological and anthropological or existentialist engagement with the world today. Second, we currently witness an increasing conflation of these types of questions with popularizations of speculative metaphysics. A look at contemporary revivals of panpsychism serves as an example to illustrate this claim. Recent support for panpsychism is visibly fueled by a science-skepticism that is resembling anti-vaccine and other pseudoscientific movements. Notably, this science-skepticism appears to draw on some arguments known from the philosophy of science. Third, I show that a number of these arguments fostering this recent science-skepticism build on a misleading image of science in which science is characterized as a cognitively unchanging practice with the primary purpose of producing knowledge output. Instead, I suggest rethinking science as the evolutionary development of *cognitive gadgets* or mental mechanisms through cultural transmission. Science is not primarily in the business of knowledge output by accumulating facts and answers; it is a cognitive activity in the pursuit of increasingly sophisticated questions for understanding the world. Such a revised image of science further offers new avenues of collaboration with the humanities in the twenty-first century.

2. BULTMANN'S PROGRAM OF DEMYTHOLOGIZING IN ITS MODERN SIGNIFICANCE

Science has transformed human understanding of the world throughout its cultural history. In turn, humanity's self-understanding was not fixed throughout its history either: how shall we think about ourselves? Bultmann's (1941) essay contains a challenge that remains as pressing to us as previous generations. What distinguishes his thinking about a clash between mythological belief and science is that his argument emphasizes the responsibility of the humanities in that conflict. Myths are not criticized merely for presenting us with an ancient cosmology in need of a scientific overhaul. Rather, the literal use of myths also stands in fundamental conflict with *our intellectual development* as *historical* and *cultural agents*:

> What is involved here (. . .) is not only the criticism that proceeds from the world picture of natural science, but also—and even more so—the criticism that grows out of our self-understanding as modern persons. (Bultmann 1989, 5)

What makes myth and science incompatible today? Myths evoke miracles, placing certain events or phenomena outside a framework of causal explanations. Mythological narratives highlight unexplained and seemingly elusive phenomena, supporting an empirically inaccessible metaphysical umbrella discussion of myths in Segal (2004). In effect, myths tell you a story to make sense of the world; however, they do not look to question our foundation of understanding of *that world*.

The value of myths is inherently social and community building. Myths provide a worldview that accounts for the second nature of humans, in Ernst Cassirer's sense (Skidelsky 2011): as humans, we are part of the natural world; yet we further question our nature in relation to our placement within this world. "Thus, myth does not want to be interpreted in *cosmological* terms but in *anthropological* terms—or better, in *existentialist* terms" (Bultmann 1989, 9; emphases added). The function of myths is to make the world in its unfamiliar forces familiar by embedding it into the experiential realm of human social life: feelings of agency, motives, meaning, and affections.

Questions surrounding our unknown and changing placement within the world are given personal certainty through social significance and meaning.

Questions about our self-understanding are tied inevitably to our familiarity with the cosmos. Why is there a universe? And why is there something rather than nothing? Who am I, what is life and why is it? Ultimately, these questions ceased being mythological. The advance of natural philosophy turned unfamiliar forces in the world from story elements with an answer into questions that had not yet been answered (and for which we may or may not find one). Natural philosophy did not replace myth by providing answers; it fundamentally transformed the human activity of asking *cosmological questions*. It offered growing insight into the cosmos' nature by advancing particular kinds of questions *as questions about the world*. Natural philosophy discharged cosmological questions from their anthropological origins.

Astrophysics, evolutionary biology, chemistry, neuroscience, and other branches of science emerged throughout our cultural history, which increasingly specialized in refining our ability to ask better questions about the nature of the world (Firestein 2012). Regardless of whether someone still feels skeptical about a full scientific answer to some of these questions, scientific explanations (involving fundamental particles in the universe, biochemical interactions between nucleotide strings, gradual changes and selective forces on organisms, information in the brain as the waxing and waning of electrical activity across cells, etc.) radically transformed society and its worldview—and continue doing so.

Still, the cultural activity embodied by myths, their substantially *anthropological* question, persists: who are *we*? This kind of investigation requires engagement different from cosmology and of an inherently existentialist character. It likewise presents us with an ongoing and, most importantly, mutable question of which science is a vital part. Science has shaped who we are and in what relation to the world we see ourselves as historical agents. Instead of "taking away" from the humanities, science has given cultural studies a new and challenging task: to study the changing image of ourselves.

> Beyond this moment of perspective, however, there is still something else to think about that I like to call the moment of "existential encounter." (. . .) historians participate with their own existence, for historical phenomena are not what they are as such—precisely as historical phenomena—without the historical subjects who understand them. (Bultmann 1989, 136)

Therefore, the value and the responsibility of the humanities are not to discuss answers to arcane metaphysical issues that elude science. It is its ongoing engagement with our own sense of agency and changing self-image as historical actors:

Events of figures in the past are not historical phenomena at all simply "in themselves," not even as parts of a causal continuum. They are historical phenomena only in their relatedness to the future, for which they have meaning and for which the present has responsibility. (Bultmann 1989, 137)

This is where Bultmann's program of demythologizing as a historical activity emerges. Engaging with metaphysical ideas and worldviews "we simply have to ask whether it really is nothing but mythology or whether the very attempt to understand it in terms of its *real intention* does not lead to the elimination of myth" (1989, 9; emphases added). Accordingly, demythologizing is not the elimination of myth. It involves the identification and analysis of mythological worldviews *as myths*, meaning a mythological picture of the world is "not to be questioned with respect to the content of its objectifying representations but with respect to the understanding of existence that expresses itself in them" (1989, 10). Conflict between myth and science arises when myths are thought to be resolutions of cosmological questions rather than being recognized for their anthropological role.

Demythologizing is not a scientific activity. However, as the remainder of this chapter shows, it requires an understanding of science. Unfortunately, such knowledge is not a given today. Consider the popular, if unhappy, rivalry of "The Two Cultures" in the British novelist C.P. Snow's (in)famous lecture of the same name. Snow (1959) highlighted a puzzling disparity in what is seen as socially relevant knowledge:

A good many times I have been present at gatherings of people who, by the standards of the traditional culture, are thought highly educated and who have with considerable gusto been expressing their incredulity at the illiteracy of scientists. Once or twice I have been provoked and have asked the company how many of them could describe the Second Law of Thermodynamics. The response was cold: it was also negative. Yet I was asking something which is the scientific equivalent of: Have you read a work of Shakespeare's? I now believe that if I had asked an even simpler question—such as, What do you mean by mass, or acceleration, which is the scientific equivalent of saying, Can you read?—not more than one in ten of the highly educated would have felt that I was speaking the same language. So the great edifice of modern physics goes up, and the majority of the cleverest people in the western world have about as much insight into it as their neolithic ancestors would have had.

Science literacy is not just of importance for the practice of science. It has become foundational to modern citizenship in the participation and development of society. Yet we continue to place an unequal balance on the kinds of knowledge which citizens require to participate in modern society. Carl

Sagan likewise emphasized the importance of science literacy as a foundation of social stability: "We live in a society absolutely dependent on science and technology and yet have cleverly arranged things so that almost no one understands science and technology. That's a clear prescription for disaster" (Sagan 2006). This is not an abstract concern about the future. We witness an increase in popular science books—routinely written by experts—that reach out to society about the dangers in current illiteracy of how our modern tools work. A case in point is *Weapons of Math Destruction* by O'Neil (2016), which demonstrates how routine applications of algorithms, when resting on an absence of understanding of these tools, increase social inequalities.

Further, we tend to be unaware of our illiteracy. A sketch by the comedian O'Brian (2012) best illustrates the point. Imagine we traveled back in time to meet the greats DaVinci, Botticelli, or Michelangelo: "We think we'd be gods with our modern wonders of technology! In reality, we are three questions away from looking like idiots." Failing to explain how our common appliances, such as computers or fridges work, quickly reveals how limited our comprehension of ordinary modern technologies is. Such existing and real disparity between our imagined knowledge and genuine understanding is called the Unread Library Effect, or "the illusion of explanatory depth." This term originates in a study by Rozenblit and Keil (2010). Participants in this study were asked to rate their confidence concerning their understanding of a toilet's workings before being prompted to provide an explanation. Following their descriptions, participants again rated their confidence in knowledge. These ratings were markedly lower the second time.

Ignorance about the extent to which we rely on "borrowed knowledge" from science grounds in a distinct lack of understanding (and appreciation) concerning the *participation* in its production. This should worry us since such a lack of understanding of science as a cognitive activity has reopened the doors to a problematic contemporary form of science-skepticism.

3. PANPSYCHISM: REPLACING ONE MYTH WITH ANOTHER MYTH

A popular joke among neuroscientists alludes to the fact that, post-retirement, some high-profile researchers started developing an interest in the mysteries of consciousness. Indeed, when invited to participate in a symposium on consciousness at an international brain science conference, a colleague of mine promptly retorted: "Well, I'm not *that* old." The investigation of consciousness has been the undisputed realm of philosophy—until now. The neural basis of consciousness became a proper (aka fundable) topic in neuroscience over the past decades: "Though still inchoate, that new science represents a true paradigm shift and a consensus that consciousness is a legitimate topic of scientific investigations" (Koch 2012, 6).

Philosophers initially responded with hostility toward this new research program, arguing that certain things lie outside the grasp of experimental science—with consciousness at the front and center. The mystery of consciousness not to be reduced to a scientific explanation is what Chalmers (1995) coined "the Hard Problem": how can physical matter give rise to subjective experience, and why? The dilemma is already implied in the semi-rhetorical framing of the question. It seems inconceivable that physical matter produces subjective consciousness. Descriptions of the interactive complexity of electrochemical processes between neurons do not seem to pair well with the experiential feeling of conscious awareness that we ascribe to ourselves.

Such alleged "explanatory gap" in the unsettled scientific nature of consciousness had some philosophers conclude that the idea of a scientific explanation of consciousness is futile (Van Gulick 2014). But this simply does not follow. It misses the point of scientific inquiry and shifts authority to fewer sound alternatives:

> *Unsettled* science is not *unsound* science. The honesty and humility of someone who is willing to tell you that they don't have all the answers, but they do have some thoughtful questions to pursue, are easy to distinguish from the charlatans who have ready answers. (Firestein 2020; emphases added)

One of these ready-made answers garnered renewed attention: panpsychism, an age-old metaphysical doctrine about consciousness that features predecessors in Ancient Greece (e.g., pre-Socratic thinkers) and the Early Modern period (e.g., Spinoza). Its popularity was at a peak during the nineteenth and early twentieth centuries (e.g., Teilhard de Chardin, Whitehead). And its lineage continued in the twentieth century (e.g., Chalmers). The attraction of panpsychism, especially for philosophers in armchairs, lies in its explicit skepticism and stark contrast with a causal-mechanistic scientific worldview (e.g., Strawson).

The central tenet of panpsychism is that consciousness is not exclusive to humans and other animals (or any living organism for that matter). Consciousness instead is introduced as a property inherent in all matter, including the fundamental elements studied in the particle accelerator at CERN. Panpsychism comes in different flavors, sometimes mimicking scientific theory (Koch 2012), sometimes openly in an appeal to personal belief (Goff 2019). Yet all these theoretical flavors come with the same aftertaste: consciousness constitutes a fundamental part of the universe (Goff et al. 2017). As a result, its conceptualization and investigation are a cosmological issue.

The scoring point for panpsychism is that it claims to have solved "the Hard Problem." Again, the proposed answer is derived from the rhetoric of the question: it seems inconceivable that physical matter can produce consciousness; therefore, consciousness must be in matter already. In other words: matter cannot and does not *create* consciousness; thus, it already *is* conscious. This results in a somewhat paradoxical belief where some philosophers "find it easy to believe that physical particles have conscious experiences" yet simultaneously "can't conceive of any way that physical brains and bodies of biological creatures could give rise to conscious experiences" (Smith 2020).

Critics noted that panpsychism isn't testable and too speculative a theory (see the debate between Goff and his critic Pigliucci 2019).[3] To this, the panpsychist response, predictably, was that some scientific ideas and concepts are also not strictly testable (Goff in response to Pigliucci 2019). So, how can you prove or disprove the metaphysical tenets of panpsychism? The point is: you cannot—moreover, you need not. Panpsychism is offered as a descriptive, not a productive worldview (meaning, it describes appearances under the lens of common intuition, although without further investigation or by offering any constructive approaches to better understanding of the phenomena it describes).

Panpsychism is noteworthy not for its view on mental matters, but for the particular philosophical argument with which it was popularized recently in the publication of *Galileo's Error: Foundations for a New Science of Consciousness* (Goff 2019). (Note the inclusion of "science" in the title of this book.) Here, its plausibility is offered not via a rigorous investigation of consciousness, but by claiming that its major competitors cannot explain consciousness either.[4] To advance his views, Goff draws on arguments resonating uncomfortably close with other science-skeptic movements, such as climate change denial and anti-vaccine advocacy (see also section 4). Such skeptics commonly appeal to the failure of normal science.

Science has failed to explain consciousness, Goff claims. The book concludes (right from the outset) that "none of this [neuroscience research] has shed any light on how the brain produces consciousness" (p. 3), and "neuroscience has thus far failed to provide even the beginnings of an explanation" (p. 5). This claim is dealt with quickly.

It is remarkably wrong. Throughout the past century, a plethora of experimental research shed new light on how to conceive the nature of consciousness (what it is, and what are the mechanisms that produce its appearance). A case in point is Sperry and Gazzaniga's split-brain research in the 1960s (Gazzaniga 2014; philosophical treatment in Schechter 2018). Here, some suggest the seemingly counterintuitive philosophical proposal that the brain may house two forms of consciousness, while others think that the hemispherectomy may lead to the emergence of two separate consciousnesses.[5]

Or, consider Goodale's 1990s work on people with visual agnosia, a condition in which people cannot consciously perceive objects but act as if they can (e.g., Patla and Goodale 1996). How much is conscious awareness and behavior shaped by subconscious information? In olfaction, for example, the conscious processing of odor qualities in mixture perception is influenced by the presence of sub-threshold components; in other words, while these components do not reach the threshold of conscious processing themselves, they affect how the qualities of components that have a high enough threshold to reach conscious awareness are perceived (Barwich 2020). Similar phenomena have been observed in other sensory modalities, including vision. Hence, to understand the nature of consciousness, it seems intellectually stifling to reflect only on what we think we experience consciously. Indeed, many experiments in modern psychology and neuroscience have opened new avenues of understanding the role and roots of conscious processes in human and other animal cognition. Some experiments on illusions of body ownership suggest that consciousness, as minimal phenomenal selfhood, connects to a sense of agency (Limanowsky 2014). Further work distinguished the neural correlates of conscious from the correlates of unconscious processing (Demertzi et al. 2019). These are only a few examples of developments in modern neuroscience that have opened radically *new perspectives and questions* to investigate the nature of consciousness. Frankly, the claim that modern research in neuroscience has not shed *any* light on how the brain produces consciousness and *failed* to provide *even the beginning* of an explanation reads oddly ill-informed considering these studies.[6]

Such mounting empirical data for Goff and other skeptics is simply beside the point. In fact, they miss the target: the inherently subjective nature of consciousness. Consciousness is defined as the intimate feeling of "what it is like," starting with what it is like to be you or me in a given moment. Consider your personal experience of the redness of red or the repulsiveness of the smell of feces. How does this subjective experience fit with a materialistic scientific worldview? Science aims to describe the world objectively from a third-person perspective. Consciousness is a subjective, first-person quality eluding the grasp of the detached neuron-stimulator. Or so the story goes.

Modern philosophy discusses this "what it is like" character of subjective experience as "qualia" (Tye 2017). What makes qualia inaccessible to science? It all starts (and apparently also ends) with Galileo Galilei:

> The argument is that Galileo opened up the way for the success of the natural sciences by restricting them to what could be measured. Sensory qualities such as sound and taste had no place in this schema and were exiled from the domain of scientific investigation. And yet sense experience is one of the elemental

aspects of the conscious mind and the basis of all scientific investigation, including that into consciousness. From the time of Galileo on, Mr. Goff says, science was structurally incapable of explaining consciousness. (Bagghini 2020)

So where did Galileo put colors, smells, and all things consciousness? He placed qualia into the soul. In response, Goff proposes to replace one myth (the soul) with another myth (conscious matter). What is explicitly ignored in this storyline is that modern cognitive psychology and neuroscience has engaged with the experiential and seemingly purely subjective aspects of mind (e.g., attention and consciousness in Graziano 2013; color vision in Chirimuuta 2015; taste and flavor in Spence et al. 2015).[7]

By way of example, take variations in smell perception: is your perception of rose or cilantro the same as mine? There now is a scientific answer to this age-old question about qualitative differences in subjective experience when it comes to the odor of things. In some cases, it involves genetics.[8] Olfaction is one of the most genetically diverse systems in the mammalian genome. Even slight genetic differences affect the qualitative perception of an odorant, that is, a volatile smelly molecule (Mainland et al. 2014; Trimmer et al. 2019). Consider cilantro, or coriander; the reason why some people dislike the aroma of cilantro and consider it soapy and pungent as opposed to fresh and green according to cilantro-likers is a genetic mutation near one of the olfactory receptor genes (Eriksson et al. 2012). Many of such puzzling phenomena in smell perception are addressed and subject to investigation in current science (Barwich 2020).

Meanwhile, so little science (or even scholarship in history and philosophy of science) features in this rhetorical ambush on modern research through one man's ahistorical reading of Galileo that one must ask: for whom is this book written—and to do what?[9] Goff does not hide the desire to model his voice after Galileo throughout the book, but his focus on the *products* of Galileo's work, rather than the human and historical *process* of Galileo's participation in science as a historical agent, causes Goff to misrepresent both Galileo and modern neuroscience.

Isn't the critical feature of science, Goff repeatedly asks, to doubt? Behind these rebel yells, panpsychism remains a markedly *passive* view. It is a view *ex negativo*. It doesn't explain or even leave you with a better understanding of what the whole philosophical point of disputing the nature of the Hard Problem is in the first place. Instead, the book *starts with an answer*, a preconceived notion of consciousness. There is no issue of doubt about what consciousness is.

Scientific theories, such as quantum mechanics, may have "undermined many of our commonsense ways of thinking about matter. However," Goff proclaims, "there is a limit to this" (a limit set as early as page 9). This limit is your private intuition about consciousness about which you need no science and which science cannot touch. Hence, nowhere is it *questioned* what consciousness is, even though our understanding of consciousness—in scientific *and* philosophical contexts—may benefit from a more critical examination of its features and nature. (The rhetoric of doubt quickly took a vow of silence here.)

Of course, some may object that consciousness is the *explanandum* for which advocates like Goff offer panpsychism as the explanans.[10] The problem with this objection is that it posits the phenomenon of consciousness as already being known in its characteristics.

Panpsychism introduces consciousness as an unmeasurable independent variable into the equation. However, just because we experience consciousness directly, do we *necessarily* know its nature? The answer is negative.

Consider an analogy with memory. As argued in Bickle and Barwich (2022), our understanding of memory, what it is and how it works, has changed fundamentally throughout the twentieth century in light of neuroscientific research. For example, memory

> can no longer be considered exclusively a higher-level cognitive phenomenon. Consider the lowly prokaryotes. In the 1970s certain bacteria (*Salmonella typhimurium*) were documented to exhibit navigation behavior along concentration gradients that resembled memory processes. (. . .) More recently, the behavior of these bacteria was shown to hinge on molecular binding mechanisms, homologous in both structure and function to NMDA receptors in LTP [Long Term Potentiation].

Additionally, memory

> has not just changed its scope as a cognitive phenomenon across different kinds of organisms. It has also changed its meaning as a cognitive process. (. . .) [T] he molecular characteristics of LTP challenge various features traditionally attributed to memory. LTP is essentially an association-sensitive molecular modification process enacted over activity-selective neural populations. Current neurobiology-inspired understanding of memory processing does not resemble the passive propositional content that philosophy traditionally has attributed to memory. Instead, molecular signal consolidation resembles the flexible mental phenomenon that contemporary psychologists have hypothesized, with memory faculties markedly malleable and processual.

In other words, our first-person experience of mental phenomena presents us with observational access to expressions of these phenomena, but not a full understanding of its nature and characteristics.

Now consider another analogy, this time with flavor perception. Flavors like strawberry, vanilla, apple, or anise are commonly considered tastes. However, we do not have a strawberry or mint receptor on our tongue and instead detect the five primary tastes of salty, sour, sweet, bitter, and umami. These flavors are an olfactory phenomenon, caused by retronasal (mouth-breathing) smell. Aromatic molecules travel from the mouth through the open pharynx to the nasal epithelium, causing us to experience a multitude of food flavors. Yet, we experience them as located in the mouth (a phenomenon called "oral referral," Spence 2016). Such examples illustrate that our direct experience is of mental phenomena is no grounds from which to infer its nature and understand its features (here: flavor as an olfactory instead of a gustatory phenomenon).

A similar argument can be made about consciousness. First-person conscious experience grants us access to observing its *expression as a mental phenomenon*. It does not provide us with a full account of what consciousness is, already given to understanding. In fact,

> [t]his quip points out the hopelessness of understanding the brain by introspection. Indeed, having a brain may be the biggest obstacle to understanding it. Not that the organ is not smart enough to understand how it works, but that the first-person experience of having a brain, what it feels like it's doing, is rarely a very good indication of what is actually happening up there. In fact, it is usually dead wrong. (Firestein, 2022)

From this perspective, panpsychism embodies a metaphysical attitude that, in (too) many of its current proposals, appears to tell us more about how a number of people *want* to think and what they want to believe about consciousness than clarify or explain something about what consciousness is. This divergence of views points us to the core of the intellectual and methodological clash between modern neuroscientific forays into consciousness and certain orthodox metaphysics of mind.

Surely the above criticisms directed cannot entirely discredit the *logical possibility* of some panpsychist claims. Notwithstanding, its *conceptual* and *physical plausibility* in the context of modern cosmological insights requires (a lot) more than mere appeals to private intuition and personal beliefs. Perhaps there could be some relevance to some of its ideas, even some of its intellectual motives and motivations. My primary concern here is that the *format of arguments* by which panpsychism has been popularized and advanced more recently is piggybacking on old-fashioned metaphysical beliefs about the nature of mind, which gain plausibility primarily from a notable degree of science illiteracy. Trying to fill-in for open questions in current science by reimposing old metaphysical beliefs, without demytholo-gizing these beliefs, leaves us with a modern myth instead of better under-standing. To be sure, not every theory needs to be truthful to be intellectually stimulating in creating new perspectives for scientific inquiry. That said, it should be stimulating in advancing new questions through such a reframed perspective.

What is therefore most problematic, in my view, is the fact that panpsy-chism doesn't raise new, better, or exciting *questions* about the nature of mat-ter or of consciousness (questions with answers that aren't already on the table of current philosophical and scientific investigations). Panpsychism notably short-changes us in deeper understanding. Say you'd proposed a theory that ascribed "consciousness" not just to living beings but to elemental physical particles like electrons. Shouldn't that theory entail and build on a sufficient understanding of electrons, their properties, and their past and present sci-entific investigation? Physicists (Hossenfelder 2020) publicly criticized the absence and conflict with modern physical theories of fundamental particles in panpsychism. Pause with this thought. What kind of understanding (both of consciousness and fundamental particles) is or will be the practical result here? The answer is a mythological image of the world—a myth that con-tains familiar elements as story-telling placeholders. Modern panpsychism presents us with a myth that does not ground in an understanding of these fundamental elements but disposes of their empirical nature, including their characteristics as scientific objects of investigation.

In effect, panpsychism claims to address a cosmological question (con-sciousness and its place in a naturalistic worldview). But it circumvents *any participation* in the scientific knowledge and its production surrounding that question. On the contrary, the material reality of science seems dismissed right from the outset (regardless of whether it is physics, biology, psychol-ogy, or neuroscience);[11] in its place, *Galileo's Error* appeals to preexisting and pre-scientific beliefs about the nature of consciousness.

Moreover, this argument for panpsychism implicitly draws on a broader science illiteracy about the brain when it appeals to "borrowed knowledge": reference to the brain using electrochemical signals seems enough to dis-miss all neuroscientific explanations as not accounting for anything related to the mystery of consciousness. But (why) is it unconceivable to explain consciousness via electrochemical signals between cells? The "illusion of explanatory depth" sends its regards: we find ourselves confronted with mythological appeals to pre-scientific intuitions about ourselves, not an argument related to the material basis of consciousness, cell processing, or electrons.

We are back at the dilemma of Bultmann's light switches.

4. REVISITING THE SCIENTIFIC IMAGE: THE COGNITIVE TURN

"What is 'Science' anyway?" asks Goff and warns of "simplistic views of science" (2019, 11). Science, in Goff's portrayal, is not "simply a matter of setting up experiments and then recording the data." (Don't just poke a few neurons; you have to be more thoughtful about the stuff that such brain poking is about!) Earlier advances of science are said to build on activities of "reimagining nature" and "dreaming up possibilities (. . .) that nobody had previously entertained" (p. 11). Besides, one imagines Goff saying, didn't Galileo do just that?

Appeals to a failure of science as the *principal* form of support for modern panpsychist theories is a dubious intellectual strategy, also known from recent pseudoscience supporters, including the anti-vaccine and climate change denial movements:

> The anti-vaccine movement claims to be based on suspicion of authority, beginning with medical authority in this case, stemming from the fraudulent data published by the now-disgraced Andrew Wakefield, an English gastroenterologist. And it's true that much of science is based on suspicion of authority. Science got its start when the likes of Galileo and Copernicus claimed that the Church, the State, even Aristotle, could no longer be trusted as authoritative sources of knowledge. (Firestein 2020)

Overthrowing authority and the radical pursuit of doubt (routinely by a single man against a homogeneous, ossified establishment) pose a provocative and captivating narrative, of course. Not by coincidence is it the main theme in many faux controversies in science, where we can witness a widening gap between the realities of scientific practice and the public perception of a research topic (e.g., olfaction in Barwich 2018). The devil is in the detail, though, because such a storyline tends to shun direct engagement with current and ongoing research in favor of battling a historical strawman. (Sounding familiar already?)

Worryingly, contemporary science-skepticism as well as certain resurrections of archaic, speculative metaphysics began to co-opt arguments from the twentieth-century philosophy of science. These philosophical arguments have highlighted the fact that scientific theories are historically replaceable. Skepticism concerning the status of our current scientific theories draws especially on Kuhn (1962), who popularized a contrast of normal science with revolutionary science via a historical narrative of paradigm shifts. Scientific theories are turned over and over (and over) to be replaced by novel frameworks based on epistemic and socially induced developments.

Additional reasons for skepticism stem from the "pessimistic meta-induction," an argument about the uncertain truth of scientific theories (Laudan 1981). On these accounts, the history of science appears to look like a graveyard of glorious failures: observations that led to earlier scientific explanations have to make way for other explanations that accommodated the same observations, which in turn got replaced by newly emerging ideas. So, given that past scientific theories were not true, what reason do we have to think that current views will stand the test of time? None whatsoever, according to several philosophers of science (review in Chakravartty 2017).

However, what constitutes the foundations of the scientific image that underlies such disputes about the foundation of scientific understanding and authority? Goff is right in one respect; we ought to ask: what *is* science? Science can be viewed as a *cultural product*, analyzed by the impact of its outputs. Alternatively, science can be conceived as *a cognitive process*, a refined strategy of interaction with the environment via instruments and collaborators.

But before we can talk about rethinking the image of science as a cognitive process, including the implications for issues surrounding Scientism and scientism, a distinction must be drawn first. To be sure, the idea to view science as a cognitive process should not be conflated with the received ideal of the scientific method™. This old epistemological image of science is still communicated in many science student handbooks and does not hold up to historical scrutiny (e.g., analysis of the non-linear progression of research and methodological developments in science in Schickore 2017). Philosophers and historians are right: theories do get repeatedly replaced by other theories, sometimes resulting in interesting intellectual challenges to our confidence in justifications of scientific knowledge. Besides, science routinely advances by failures (Wimsatt 1987; Firestein 2015; Barwich 2019; Schickore 2021). Consequently, by ambushing science in terms of its fallibility, advocates of science-skepticism and pseudoscience pick a free ride mimicking an idealized and, moreover, outdated version of the image of science as the pursuit of Truth™. This image, regardless of its nobility, cultivates a misleading, even harmful caricature of science today. It resembles science as much as scented candles labeled "ocean fresh" smell of the Atlantic.

What does talk about "science as a cognitive process" entail? Consider the metaphors we use to talk about science (list adopted and quoted from Firestein 2013):

- the puzzle metaphor (the analogy of scientific activities with puzzle solving also appeared in Kuhn (1962))[12]: "scientists are patiently putting the pieces of a puzzle together to reveal some grand scheme or another"
- the onion (used, for instance, by Feynman (2005)): "science is busy unraveling things the way you unravel the peels of an onion. So peel by peel,

you take away the layers of the onion to get at some fundamental kernel of truth"

- the iceberg (in Firestein 2013): the idea "that we only see the tip of the iceberg, but underneath is where most of the iceberg is hidden"

These metaphors convey a sufficiently similar understanding of science: science accumulates facts and pursues the sampling of knowledge to arrive at an exhaustive and detailed picture of the cosmos. However, neuroscientist Firestein (2012, 2013, 2015, 2020) has cautioned that this ends up a misconception. The perpetuation of this image may even cause harm to science education and communication, especially in modern society. Because what these metaphors don't account for is the process of science and the nature of participation in its activities. In other words, it centers the product of science over the participation in the process of its production. However, it is questionable to what extent one can evaluate the nature of scientific outputs properly without a meaningful understanding concerning their production.

However, public and even some philosophy discourse continue to emphasize the systematic justification of models over their generation as the central characteristic of science (e.g., Hoyningen-Huene 2008). But the business of science is essentially *exploratory*, which can significantly diverge from the predominant philosophical interest in its justification (Schickore 2016; Gelfert et al. 2021). Orthodox philosophical attention to the justification of knowledge gave rise to a severely distorted understanding of science in action that fails to handle unsettled and ongoing science (Barwich 2018).

Recently, for example, the public got a taste of real science in light of the fast changeability of COVID-19 models (Stegenga 2020). What may have resembled some semi-disorganized bumbling in the dark was a glimpse of science in action: namely, when you do not know the answer yet and are on your way to figuring it out. It felt disorienting only because our public and philosophical narratives about how science works have been built via reconstructions of historical examples, a science narrative with already known answers. But science is not ready-made or finished. "This is like forgetting that the end product of apple trees is not apples—it's more apple trees" (Dennett 1993, 9; original quote on the nature of perception).

Likewise, the purpose of asking questions with science is not accumulating facts. It is furthering questions. Or in Firestein's (2013) words:

> The facts are important. You have to know a lot of stuff to be a scientist. That's true. But knowing a lot of stuff doesn't make you a scientist. You need to know a lot of stuff to be a lawyer or an accountant or an electrician or a carpenter. But

in science, knowing a lot of stuff is not the point. Knowing a lot of stuff is there to help you get to more ignorance.

Developing better and more sophisticated questions requires insight and training about knowing what to ask, which involves a sufficient level of "informed ignorance" about what is unknown and where our understanding benefits from further investigation. Science is not a puzzle in need of completion. It better compares to ripples on a pond: in its center is what we currently know, while its circumference represents the ignorance about what we do not yet know (Firestein 2012, 2013).

On this account (figure 8.1), the growth of knowledge results in an expansion of ripples in parallel with a growth of their circumference of simultaneously expanding ignorance. What is changing, though, is the *sophistication* of ignorance.

Scientific theories according to this image of science don't get lost and replaced entirely. Instead, they leave traces in the sediments of the circle, such that an archeologist of knowledge can uncover an evolutionary development of explanations (Foucault 2012[1969])—including the branching out of varieties of knowledge, the exaptation of older explanations to adopt new functions in other (if not even the same) scientific frameworks, and so on. Science viewed in its epistemic evolution, or evolutionary epistemology (Campbell 1974), further reveals that explanations are not replaced equally. Some explanations become more challenging to replace than others, and some increasingly so

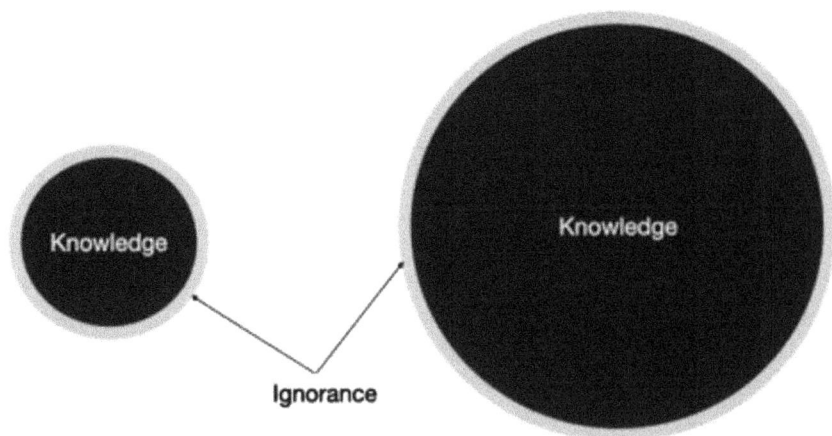

Figure 8.1 Idea adopted from Chang 2012, 256: Alternative conception of scientific progress as the cultivation and sophistication of ignorance. With the growth of knowledge also expands our ignorance concerning our awareness of the things yet unknown (the open questions).

(Deutsch 2011). These constitute good or better explanations. Are they never to be replaced? Of course not, as nothing in any evolutionary process remains fixed and unchanging. In effect, this understanding constitutes a paradigm shift from a Newtonian toward a Darwinian model of science.

Such an evolutionary model of scientific development has also entered the philosophy and sociology of science. What is considered evolving here are the scientific theories, social contexts, and instruments by which scientists make observations. In this context, interactions between twentieth-century epistemological and sociological programs have been of mutual benefit, so much so that sharp boundaries between these two fields became scrutinized and, sometimes, rejected (Longino 1990; Solomon 2007). Many influential works in the history of the philosophy of science have aimed to integrate historical developments and social data into their frameworks, such as Kuhn (1962), Lakatos (1968), Laudan (1978). (Kuhn in his later works also promoted an evolutionary model of scientific growth (Wray 2011).[13]) Still, these frameworks explicitly favored the promotion of meta-methodological principles to parse and explain conceptual changes and to *justify choices* between scientific models as representations of knowledge. Simultaneously, sociologists disputed the priority of such meta-methodological tenets; they aimed to *explore the construction* of the outputs of science—knowledge and technology—as cultural products. In a way, both fields have remained true to their disciplinary settings: they both continued to center on the outputs of science as the primary target of analysis (detailed review of these disciplinary developments in Giere 1988). Overall, scientific knowledge—and its representation as theories, models, ideas—was seen as undergoing constant evolution.

Yet, what if we further expand our views on the evolutionary development undergirding the development of science? What if we further direct attention to the *process of participating in scientific knowledge production* itself? Specifically, we want to pay attention to the development and expansion of the material, epistemological, and *cognitive* dimension of scientific knowledge production.

Let's first consider the material expansion of science. Science, quite literally, facilitates *material growth* of our universe. Consider advances in materials science, where chemical and other engineers now create materials with striking and sometimes new properties by playing and reconfiguring their structural composition (Ball 1999). Moreover, consider synthetic chemistry which enabled us to create new fragrant compounds that have never existed on earth, often with smells formerly unknown in nature. In virtue of this material expansion, we now have learned that smell is the only sense for which we can artificially create an entirely new material stimulus (Barwich 2020). How's that for asking new and better questions about the world and our perceptions of it!

Second, there is the epistemological expansion of science in its historical trajectory. A lot has happened since Galileo. Objectivity, a hallmark of science in modern views, emerged only in the nineteenth century with increasing use and emphasis on visualizations in scientific research, its communication, and comparability (as argued by Daston and Galison 2007). Objectivity is a remarkable historical product, an achievement of science as a developing cognitive activity, not an epistemic given. The same case holds for the methodological requirements of standardization and coordination, for example, in the emergence of modern scientific writing (Schickore 2017).

Not only the products but also our participation in their production, meaning science as a cognitive *activity*, have altered our material interactions with the world—including (yet not to be concluded with) the growth of knowledge and the development of tools to do so. This activity, with its tool uses, ultimately transformed our environment.

It has likewise shaped us. For example, it changed how we interact with each other and our behavior in time. Consider the invention of clocks and the establishment of a standardized global time; these developments ultimately shaped various human activities, such as sleep rhythms and ordinary understanding of sleep, social standards for coordinating meetings, and punctuality, as well as more abstract conceptualizations of temporal notions such as synchronicity and simultaneity (Ramirez 2020).

Meanwhile, we may further need to look at scientific activities, like modeling and instrumentally guided observation, as facilitating the growth and scaffolding of cognitive mechanisms. Multiple proposals have highlighted the benefits and even argued for the necessity of this approach for modern science studies. In *Explaining Science: A Cognitive Approach* (1988) Giere argued that the philosophical study of science benefits from drawing on recent theories in cognitive science. Other prominent proposals advocating a cognitive approach to science include Thagard's (2012) computational program and Churchland's (1995) connectionist framework. And, notably, Nersessian has been at the forefront of applying theories of distributed cognition directly to concept formation in specific modeling practices (Liu et al. 2008).

In addition to the growth and expansion of materials and the epistemological dimensions of science, we thus may also want to look at the possibility of an expansion of the cognitive mechanisms through scientific tools and skillsets themselves. This idea links to recent theories combining cognitive science with cultural evolution (Heyes 2018). Approaching science as an umbrella for cognitively developing and further evolving processes from the viewpoint of cultural transmission indeed presents us with a radically new outlook:

> We humans have created not just physical machines—such as pulleys, traps, carts, and internal combustion engines—but also mental machines; mechanisms

of thought, embodied in our nervous systems, that enable our minds to go further, faster, and in different directions than the minds of other animals. (Heyes 2018, 1)

The historical development of science has facilitated multiple distinguished and novel ways of thinking—for humans to ask better and more sophisticated questions about the reality of their surroundings. Science is an activity that develops as a set of *cognitive gadgets* or mental mechanisms via its cultural transmission, which includes the development and teaching of reasoning methods and tool use as scaffolds by which cognitive skills can develop and manifest (this developmental notion of scaffolds is adopted from Caporael et al. 2014).

Of course, much more investigation is needed regarding the nature of cognitive development of mental mechanisms via cultural evolution in the practice of science. The point here is to highlight that scientific thinking cannot be analyzed somewhat ahistorically as an unchanging cognitive procedure from Galileo to today that's birthed by the scientific revolution without affecting human cognition and the mental mechanisms with which we engage with the world (as it is, for instance, propagated in Goff 2019). The activity of science and the participation by cognitive agents within its practices consists of much more than culturally inherited information that gets passively taught and rethought. Essentially, science is demarcated by the active development of an *evolving cognitive skillset*.

In this light, philosophical quibble about the failure of (neuro)science to explain consciousness falls short on several accounts. It first and foremost fails to understand what science actually *is* and how it facilitates the apperception of the world. It also fails to engage with the ways in which human cognition works, and how to investigate the phenomena we are interested in understanding, including consciousness. In other words: current science does not fail in "dreaming up possibilities," as science isn't about *imagining answers*. Science is about cultivating curiosity about how the world works *beyond* preconceived philosophical intuitions and situated beliefs, which it does by developing more sophisticated and sometimes new questions. Georg Cantor revolutionized mathematical thinking by asking whether there are different infinities. Albert Einstein redefined our thinking about time and space by reimagining the question of what simultaneity is. Charles Darwin shifted biological perspectives about development and inheritance by introducing the idea of probability into the organic world. Patricia Churchland dared to question how preconceived philosophical concepts of mind, including consciousness, change once reviewed in light of recent insights into the brain (instead of exercising scholastic metaphysics). Now, what is this radical line of inquiry about either consciousness or matter coming from this account of panpsychism? We are left with a story that tells us everything is sort of conscious.

5. OUTLOOK: SCIENCE AND THE RESPONSIBILITIES OF THE HUMANITIES IN THE TWENTY-FIRST CENTURY

Bultmann's question must be asked anew and on repeat, not only because of its specific answer and treatment of myths and science, rather, it reminds us of the vital responsibility of the humanities: as a collaborator, not a competitor, of the sciences in the twenty-first century.

The danger to the humanities is not a takeover of Scientism. The real threat is a form of irrelevance that links to an inability to ask better questions about ourselves as thinking beings and historical citizens *participating* in this world. This participation ultimately must also include the cultural evolution in our cognitive and material engagement with the world—through science.

Returning to Bultmann's Program, the responsibility of the humanities in modern times is to critically analyze the extent to which the use of metaphysical ideas builds on mythological stories, and to demythologize them by analyzing their value and intention in modern times beyond these mythological origins. Instead of resurrecting old metaphysical frameworks, attempting to replace one myth with another, modern metaphysical engagement with science and mind ought to question how our understanding of mind and matter (still) roots in certain metaphysical dogmas. And here, we find an opportunity to revisit Bultmann's answer. While humanities have a fundamentally existentialist and anthropological task that does not posit them as passive spectators in the pursuit of cosmological questions by science:

> This is also a chance to slingshot naturalist philosophy into the twenty-first century, and break down the silos of institutionalized disciplinarity that are neatly dividing philosophical from neuroscientific inquiry. Philosophy must abandon the attitude that its ideas and debates are timeless and, in their application, independent of empirical developments. Neuroscience yields new philosophical questions and angles from which to revisit historically grown, deeply engrained intuitions about the mind and its creation, structure, and environment. We need to adapt our methods to the puzzle, not the other way around. (Barwich 2020, 311)

For the puzzle of consciousness, these methods involve neuroscience *in tandem with* philosophy. A most fruitful and productive philosophical approach is to rethink which conceptions of mind and matter in modern science constitute intellectual artifacts and how to conceive of alternatives *in light of* (not despite) modern insights.[14]

Discussions about scientism thus reveal a deeper methodologically grounded divergence in modern views about the role and skills of the humanities. As this chapter has foregrounded, a central responsibility of the humanities in the twenty-first century—in addition to other roles like engaging with existentialist questions—is to develop and pose questions about the conceptual foundations of cosmological questions, ultimately and inevitably *in collaboration with* science.[15]

ACKNOWLEDGMENTS

I thank Moti Mizrahi for his invitation to this volume, Matthew Cobb and an anonymous reviewer for their constructive suggestions that helped clarify my argument in this chapter, and Matthew Rodriguez for his editorial suggestions.

NOTES

1. I thank an anonymous reviewer for this clarification.

2. For clarity, I will continue to capitalize Scientism when discussing the former view, and leaving scientism uncapitalized when discussing the latter.

3. Pigliucci's views on scientism and the issue with demarcating science from pseudoscience in Pigliucci and Boudry (2013).

4. The historical option of Dualism (ala Descartes) claims the existence of one thing too many that it cannot prove (*res cogitans*). In contrast, science's modern materialism posits one thing too few: it wants to abolish consciousness. (Frankly, this constitutes an incorrect portrayal of the latter; see also footnote 14.)

5. Thanks to Matthew Cobb for his comments on this issue.

6. Indeed, this lack of engagement with modern research into mind and brain comes close to proclaiming that physics has failed to explain the nature of the ether.

7. And while mental processes such as perception, memory, attention are sometimes nonchalantly referred to as "the Easy Problem," as opposed to "the Hard Problem" of consciousness itself, the seemingly subjective experience of perceptual qualities constitutes the bulwark of arguments against materialistic explanations of mind (a logical deconstruction of popular qualia-based philosophical arguments in Churchland 1986).

8. In other cases, it involves familiarity with smell, age, sex, and other biological and psychological factors (see Barwich 2020).

9. To be sure, and adjusting the impression conveyed in my polemics, I encountered Goff as a pleasant person in personal conversation. Meanwhile, his writing is riddled with insufferable polemics and cheap shots at intellectual opponents (including the dating life of the Churchlands)—which invite some treatment in kind.

10. I thank a reviewer for bringing this up.

11. Goff responded to my criticism (sent via email in personal communication) when I had finished this chapter with a public blog post (2020), in which he stated that he is not opposed to science. I do think this ought to be rendered more visible in research also to mirror such statement, and I am looking forward to more scientifically stimulating discussions surrounding his views on panpsychism.

12. I thank a reviewer for highlighting this.

13. Another notable point brought into my attention by a reviewer.

14. Indeed, contrary to Goff's description in *Galileo's Error*, this is what the program of eliminative materialism by Patricia and Paul Churchland really is about (e.g., McCauley 1996; Churchland 2013).

15. Many humanities scholars and scientists have taken on this challenge (e.g., Churchland 1986, 1996; Silva and Bickle 2009; Chang 2012; Firestein 2012; Heyes 2018; Ramirez 2020; Barwich 2020). It is time to highlight this new work rather than getting trapped in a dualism that artificially divides research into the roles of two separate cultures.

REFERENCES

Bagghini, J. (2020). "'Galileo's Error' Review: Of Mind and Matter." *The Wall Street Journal*, January 3. https://www.wsj.com/articles/galileos-error-review-of-mind-and-matter-11578067406 (last accessed October 18, 2020).

Ball, P. (1999). *Made to Measure: New Materials for the 21st Century.* Princeton, NJ: Princeton University Press.

Barwich, A. S. (2018). "How To Be Rational About Empirical Success in Ongoing Science: The Case of the Quantum Nose and Its Critics." *Studies in History and Philosophy of Science Part A*, 69: 40–51.

Barwich, A. S. (2020). *Smellosophy: What the Nose Tells the Mind.* Cambridge: Harvard University Press.

Bickle, J., and A. S. Barwich. (2022). "Chapter 3: An Introduction to Molecular and Cellular Cognition." In *Mind, Cognition, and Neuroscience: A Philosophical Introduction*, edited by B. D. Young and C. Dicey Jennings. London: Routledge.

Bradie, M., and W. Harms. (2020). *Evolutionary Epistemology.* Edited by M. Zalta. Winter Edition. https://plato.stanford.edu/entries/epistemology-evolutionary/.

Bultmann, R. (1989). *New Testament Mythology and Other Basic Writings.* Philadelphia: Fortress Press.

Burnett, T. (2014). "What Is Scientism?" *American Association for the Advancement of Science.* https://www.aaas.org/programs/dialogue-science-ethics-and-religion/what-scientism; https://leapsmag.com/pseudoscience-is-rampant-how-not-to-fall-for-it/ (last accessed October 18, 2020).

Campbell, D. (1974). "Evolutionary Epistemology." In *The Philosophy of Karl R. Popper*, edited by P. A. Schilpp, 412–463. LaSalle, IL: Open Court.

Caporael, L. R., J. R. Griesemer, and W. C. Wimsatt (Eds). (2014). *Developing Scaffolds in Evolution, Culture, and Cognition* (Vol. 17). Cambridge: MIT Press.

Chalmers, D. J. (1995). "Facing Up to the Problem of Consciousness." *Journal of Consciousness Studies*, 2(3): 200–219.

Chang, H. (2012). *Is Water H₂O? Evidence, Realism and Pluralism* (Vol. 293). Dordrecht: Springer Science & Business Media.

Chirimuuta, M. (2015). *Outside Color: Perceptual Science and the Puzzle of Color in Philosophy*. Cambridge: MIT Press.

Churchland, P. S. (1989). *Neurophilosophy: Toward a Unified Science of the Mind-Brain*. Cambridge: MIT Press.

Churchland, P. M. (1996). *The Engine of Reason, the Seat of the Soul: A Philosophical Journey into the Brain*. Cambridge: MIT Press.

Churchland, P. M. (2013). *Matter and Consciousness*. Cambridge: MIT Press.

Cobb, M. (2020). *The Idea of the Brain: The Past and Future of Neuroscience*. New York: Basic Books.

Daston, Loraine, and Peter Galison. (2008). *Objectivity*. Princeton, NJ: Princeton University Press.

Demertzi, A., E. Tagliazucchi, S. Dehaene, G. Deco, P. Barttfeld, F. Raimondo, C. Martial, D. Fernández-Espejo, B. Rohaut, H. U. Voss, and N. D. Schiff. (2019). "Human Consciousness is Supported by Dynamic Complex Patterns of Brain Signal Coordination." *Science Advances*, 5(2): eaat7603.

Dennett, D. C. (1993). *Consciousness Explained*. London: Penguin UK.

Deutsch, D. (2011). *The Beginning of Infinity: Explanations That Transform the World*. London: Penguin UK.

Eriksson, N., S. Wu, C. B. Do, A. K. Kiefer, J. Y. Tung, J. L. Mountain, D. A. Hinds, and U. Francke. (2012). "A Genetic Variant Near Olfactory Receptor Genes Influences Cilantro Preference." *Flavour*, 1(1): 1–7.

Feynman, R. P. (2005). *The Pleasure of Finding Things Out: The Best Short Works of Richard P. Feynman*. New York: Basic Books.

Firestein, S. (2012). *Ignorance: How It Drives Science*. New York: Oxford University Press.

Firestein, S. (2013). "The Pursuit of Ignorance." *TED*. https://www.ted.com/speakers /stuart_firestein (last accessed October 18, 2020).

Firestein, S. (2015). *Failure: Why Science is So Successful*. New York: Oxford University Press.

Firestein, S. (2020). "Pseudoscience Is Rampant: How Not to Fall for It." *Leapsmag*. https://leapsmag.com/pseudoscience-is-rampant-how-not-to-fall-for-it/ (last accessed October 18, 2020).

Firestein, S. (Forthcoming). "Preface." In: *The Tools of Neuroscience Experiment: Philosophical and Scientific Perspectives*, edited by J. Bickle, C. Craver, and A. S. Barwich. London: Routledge.

Foucault, M. (2012 [1969]). *The Archeology of Knowledge*. New York: Knopf Doubleday Publishing Group.

Gazzaniga, M. S. (2014). "The Split-Brain: Rooting Consciousness in Biology." *Proceedings of the National Academy of Sciences*, 111(51): 18093–18094.

Gelfert, A., G. Fisher, and F. Steinle (Eds). (2021). SI: Exploratory Models and Exploratory Modelling in Science. In *Perspectives on Science*. Cambridge: MIT Press.

Giere, R. (1988). *Explaining Science: A Cognitive Approach*. Chicago, IL: Chicago University Press.

Goff, P. (2019). *Galileo's Error: Foundations for a New Science of Consciousness*. New York: Pantheon.

Goff, P. (2020). *Conscience and Consciousness*. https://conscienceandconsciousness .com/2020/10/29/is-it-the-job-of-science-or-philosophy-to-account-for-conscious-ness/ (published October 29, 2020; last accessed July 20, 2021).

Goff, P., and M. Pigliucci. (2019). "Panpsychism and the Science of Consciousness." *WikiLetter*. https://letter.wiki/conversation/277 (last accessed October 18, 2020).

Goff, P., W. Seager, and S. Allen-Hermanson. (2017). "Panpsychism." In *Stanford Encyclopedia of Philosophy*, edited by M. Zalta. Spring Edition. https://plato.stan-ford.edu/entries/panpsychism/.

Graziano, M. S. (2013). *Consciousness and the Social Brain*. New York: Oxford University Press.

Heyes, C. (2018). *Cognitive Gadgets: The Cultural Evolution of Thinking*. Cambridge: Harvard University Press.

Hossenfelder, Sabine. (2020). "Electrons Don't Think." *Nautilus*. http://nautil.us/blog /electrons-dont-think (last accessed October 18, 2020).

Hoyningen-Huene, P. (2008). *Systematicity: The Nature of Science*. New York: Oxford University Press.

Hutchinson, I. (2011). *Monopolizing Knowledge: A Scientist Refutes Religion-Denying, Reason-Destroying Scientism*. Belmont, MA: Fias Publishing.

Koch, Christof. (2012). *Consciousness: Confessions of a Reductionist*. Cambridge: MIT Press.

Kuhn, T. S. (1962). *The Structure of Scientific Revolutions*. Chicago, IL: University of Chicago Press.

Labron, T. (2011). *Bultmann Unlocked*. London: A&C Black.

Ladyman, J. (2018). "Scientism with a Humane Face." In *Scientism: Prospects and Problems*, edited by J. de Ridder, R. Peels, and R. van Woudenberg. New York: Oxford University Press.

Laudan, L. (1978). *Progress and Its Problems: Towards a Theory of Scientific Growth* (Vol. 282). Berkeley, CA: University of California Press.

Laudan, L. (1981). "A Confutation of Convergent Realism." *Philosophy of Science*, 48(1): 19–49.

Limanowski, J. (2014). "What Can Body Ownership Illusions Tell Us About Minimal Phenomenal Selfhood?" *Frontiers in Human Neuroscience*, 8: 946.

Liu, Z., N. Nersessian, and J. Stasko. 2008. "Distributed Cognition as a Theoretical Framework for Information Visualization." *IEEE Transactions on Visualization and Computer Graphics*, 14(6): 1173–1180.

Longino, H. E. (1990). *Science as Social Knowledge: Values and Objectivity in Scientific Inquiry*. Princeton, NJ: Princeton University Press.

Mainland, J. D., A. Keller, Y. R. Li, T. Zhou, C. Trimmer, L. L. Snyder, A. H. Moberly, K. A. Adipietro, W. L. L. Liu, H. Zhuang, and S. Zhan. (2014). "The Missense of Smell: Functional Variability in the Human Odorant Receptor Repertoire." *Nature Neuroscience*, 17(1): 114–120.

McCauley, R. N. (Ed.). (1996). *The Churchlands and Their Critics*. Cambridge, MA: Blackwell.

O'Brian, D. (2012). "Craig Dealer Live." *Open Mike Productions*. https://www.imdb.com/title/tt2529272/ (last accessed October 18, 2020).

O'Neil, C. (2016). *Weapons of Math Destruction: How Big Data Increases Inequality and Threatens Democracy*. New York: Broadway Books.

Patla, A. E., and M. A. Goodale. (1996). "Obstacle Avoidance During Locomotion is Unaffected in a Patient With Visual Form Agnosia." *NeuroReport*, 8(1): 165–168.

Pigliucci, M. (2018). "The Problem with Scientism." *Blog of the APA*. https://blog.apaonline.org/2018/01/25/the-problem-with-scientism/ (last accessed October 18, 2020).

Pigliucci, M., and M. Boudry (Eds). (2013). *Philosophy of Pseudoscience: Reconsidering the Demarcation Problem*. Chicago, IL: University of Chicago Press.

Ramirez, A. (2020). *The Alchemy of Us: How Humans and Matter Transformed One Another*. Cambridge: MIT Press.

Rozenblit, L., and F. Keil. (2002). "The Misunderstood Limits of Folk Science: An Illusion of Explanatory Depth." *Cognitive science*, 26(5): 521–562.

Sagan, C. (2006). *Conversations With Carl Sagan*. Mississippi: University Press of Mississippi.

Schechter, E. (2018). *Self-Consciousness and "Split" Brains: The Minds' I*. New York: Oxford University Press.

Schickore, J. (2016). "'Exploratory Experimentation' as a Probe into the Relation Between Historiography and Philosophy of Science." *Studies in History and Philosophy of Science Part A*, 55: 20–26.

Schickore, J. (2017). *About Method: Experimenters, Snake Venom, and the History of Writing Scientifically*. Chicago, IL: University of Chicago Press.

Schickore, J. (2021). "Is "Failing Well" a Sign of Scientific Virtue?" In *Virtues und Epistemology*, edited by Thomas Stapleford and Emanuele Ratti. New York: Oxford University Press.

Segal, R. A. (2004). *Myth: A Very Short Introduction*. New York: Oxford University Press.

Silva, A. J., and J. Bickle (Eds). (2009). "The Science of Research and the Search for Molecular Mechanisms of Cognitive Functions." In *The Oxford Handbook of Philosophy and Neuroscience*. DOI: 10.1093/oxfordhb/9780195304787.003.0005.

Skidelsky, E. (2011). *Ernst Cassirer: The Last Philosopher of Culture*. Princeton, NJ: Princeton University Press.

Smith, B. C. (2020). *Tweet*, January 18, 7.15 pm. https://twitter.com/smithbarryc/status/1218688332420481025?s=20 (last accessed January 19, 2020).

Snow, C. P. (1959). *The Two Cultures and the Scientific Revolution.* Cambridge: Cambridge University Press.

Solomon, M. (2007). *Social Empiricism.* Cambridge: MIT Press.

Sorrell, T. (1991). *Scientism: Philosophy and the Infatuation with Science.* London: Routledge.

Spence, C. (2016). "Oral Referral: On the Mislocalization of Odours to the Mouth." *Food Quality and Preference*, 50: 117–128.

Spence, C., B. Smith, and M. Auvray. (2015). "Confusing Tastes and Flavours" (Chapter 10). In *Perception and Its Modalities*, edited by Dustin Stokes, Mohan Matthen, and Stephen Biggs, 247--274. New York: Oxford University Press.

Stegenga, J. (2020). "Fast Science and Philosophy of Science." *BJPS|Auxiliary Hypotheses.* http://www.thebsps.org/auxhyp/fast-science-stegenga/ (last accessed July 29, 2021).

Thagard, P. (2012). *The Cognitive Science of Science Explanation, Discovery, and Conceptual Change.* Cambridge: MIT Press.

Trimmer, C., Andreas Keller, Nicolle R. Murphy, Lindsey L. Snyder, Jason R. Willer, M. H. Nagai, Nicholas Katsanis, Leslie B. Vosshall, Hiroaki Matsunami, and Joel D. Mainland. (2019). "Genetic Variation Across the Human Olfactory Receptor Repertoire Alters Odor Perception." *Proceedings of the National Academy of Sciences*, 116(19): 9475–9480.

Tye, M. (2017). "Qualia." In *Stanford Encyclopedia of Philosophy*, edited by M. Zalta. Winter Edition. https://plato.stanford.edu/entries/qualia/.

Wimsatt, W. C. (1987). "False Models as Means to Truer Theories" (Chapter 2). In *Neutral Models in Biology*, edited by Matthew H. Nitecki and Antoni Hoffman, 23–55. New York: Oxford University Press.

Wray, K. B. (2011). *Kuhn's Evolutionary Social Epistemology.* Cambridge: Cambridge University Press.

Chapter 9

Whither Academic Philosophy?

Moti Mizrahi

In a "Big Think" YouTube video to promote his book, *Intuition Pumps and Other Tools for Thinking* (2013), Daniel Dennett said the following:

> One of the reasons I wrote this book is because, oddly enough, philosophers who are famous, notorious for being navel-gazers, for being reflective, I think, in fact, *philosophers are often remarkably unreflective about their own methodology.* I wanted to draw attention to how philosophers actually go about their business and get them thinking more self-consciously about the tools they use and how they use them. (emphases added)[1]

Arguably, this was false when Dennett said it back in 2013, but it is demonstrably false now. For one thing, the academic journal *Metaphilosophy* has been publishing papers "about philosophy and particular schools, *methods* or fields of philosophy" (emphasis added) since its first issue back in 1970.[2] For another, the advent of experimental philosophy in the early 2000s has ushered in an era in which academic philosophers pay careful attention to and have great interest in examining "the tools they use and how they use them." As Dolcini (2016, 102) puts it, "The advent of experimental philosophy, around fifteen years ago, revitalized the discussion about a major meta-philosophical issue: what are the proper methods, aims and ambitions of philosophy?" Similarly, according to Aberdein and Inglis (2019, 3), "The advent of experimental philosophy has not been without controversy and has *provoked a salutary debate on the proper method of philosophical enquiry*" (emphases added).[3]

Now, some proponents of scientism, especially of the weak and moderate varieties of scientism,[4] call on academic philosophers to adopt the empirical methods of the sciences. For example, Buckwalter and Turri (2018) argue for what they call "moderate scientism." Moderate scientism is the view that the empirical methods of the sciences can help test hypotheses and answer questions in non-scientific disciplines. In particular, Buckwalter and Turri argue for moderate scientism in academic philosophy, since the application of methods from the social sciences in philosophy (also known as "experimental philosophy") has been quite successful. As they put it, "Experimental, observational, and statistical techniques have significantly contributed to research in epistemology, action theory, ethics, philosophy of language, and philosophy of mind" (Buckwalter and Turri 2018, 282). Similarly, I have argued that the introduction of methods from data science into philosophy of logic in particular (Mizrahi 2019), and philosophy in general (Mizrahi 2018b, 48), might bring to academic philosophy the sort of success enjoyed by the sciences. And Van De Poel (2020, 231–244) argues that, if they want to make academic philosophy "societally relevant," academic philosophers need to incorporate the synthetic methods of designers as well as techniques of experimentation into philosophical inquiry.

On the other hand, some opponents of scientism suggest that making academic philosophy more scientific, through the introduction of the empirical methods of the sciences into philosophy, amounts to simply making philosophy more interdisciplinary. Here is an example:

> The introduction of scientific methods into philosophical inquiry is an example of interdisciplinarity, not scientism. Other examples of interdisciplinarity in philosophy are subdisciplines such as philosophy of physics and philosophy of chemistry. (Bishop 2019, 46)[5]

There are several problems here. First, the introduction of scientific methods into philosophical inquiry is not supposed to be an example of scientism per se. Second, when proponents of scientism call on academic philosophers to adopt the empirical methods of the sciences, they are not merely trying to make academic philosophy more interdisciplinary, and philosophy of physics and philosophy of chemistry are not good examples of interdisciplinarity in academic philosophy. Finally, opponents of scientism would probably protest against the borrowing and adoption of the empirical methods of sciences by academic philosophers for they would consider that "scientific expansionism" (Stenmark 2004) or "scientific imperialism" (Kitcher 2012), which is not the same as scientism. In what follows, I discuss these problems with Bishop's (2019) claims in order.

1. SCIENTISM AND THE CALL TO MAKE PHILOSOPHY MORE SCIENTIFIC

First, the introduction of scientific methods into philosophical inquiry is not supposed to be an example of scientism per se. The former is a policy of action, whereas the latter is a *thesis* (or, more precisely, a set of various theses) about science in comparison to other ways of knowing. In its epistemological form, scientism is the thesis that scientific knowledge (or some other epistemic good, such as justified belief) is superior to non-scientific knowledge (or some other epistemic good), either because scientific knowledge is *better* than non-scientific knowledge or because *only* scientific knowledge counts as knowledge. This epistemological thesis says nothing about introducing any methods into academic philosophy specifically. Similarly, in its methodological form, scientism is the thesis that scientific methods are superior to non-scientific methods, either because scientific methods are *better* than non-scientific methods or because *only* scientific methods produce knowledge (or some other epistemic good). Again, this methodological thesis says nothing about introducing any methods into academic philosophy specifically. Finally, in its metaphysical form, scientism is the thesis that science is superior to non-science as a guide to the nature of reality, because science is either the *best* or the *only* guide to the nature of reality.[6] Again, this metaphysical thesis says nothing about introducing any methods into academic philosophy specifically.

Rather, contrary to what Bishop (2019) seems to think, the call to introduce the empirical methods of the sciences into academic philosophy is a *corollary* of scientism understood as a methodological thesis. More specifically, a strong version of methodological scientism along internal/academic lines states that the empirical methods of the sciences are the *only* methods that give us knowledge about the world and ourselves.[7] On this view, then, if non-scientific disciplines are to produce knowledge, they *must* employ the empirical methods of the sciences as well. This is a corollary of a strong version of methodological scientism along internal/academic lines. It is not what the view itself states. Likewise, a weak version of methodological scientism along internal/academic lines states that the empirical methods of the sciences are the *best* methods that give us knowledge about the world and ourselves.[8] On this view, then, if non-scientific disciplines are to produce knowledge, they *better* employ the empirical methods of the sciences as well. Again, this is a corollary of a weak version of methodological scientism along internal/ academic lines. It is not what the view itself states. Clearly, then, methodological scientism, which is the view that scientific methods are superior to non-scientific methods, and the call for academic philosophers to adopt the empirical methods of the sciences are not one and the same. Nor is the latter an example of the former, as Bishop (2019) seems to think. Instead, the call to introduce the empirical methods of the sciences into academic philosophy is a *corollary* of methodological scientism.

The same point applies to the view Buckwalter and Turri (2018) call "moderate scientism," which is the view that the empirical methods of the sciences can help test hypotheses and answer questions in non-scientific disciplines. If the empirical methods of the sciences can help test hypotheses and answer questions in non-scientific disciplines, and practitioners of those non-scientific disciplines want to test hypotheses and answer questions in their disciplines, then they should probably use the empirical methods of the sciences. Again, this is a corollary of moderate scientism, not a statement of the view itself. Applied to academic philosophy, then, academic philosophers should probably adopt the empirical methods of the sciences, if they want to test hypotheses and answer questions in philosophy, for the empirical methods of the sciences can help philosophers attain their disciplinary goals, namely, test hypotheses and answer questions in philosophy. Clearly, then, moderate scientism, which is the view that scientific methods can help non-scientists answer questions in their non-scientific disciplines and the call for academic philosophers to adopt the empirical methods of the sciences are not one and the same. Nor is the latter an example of the former, as Bishop (2019) might think. Instead, the call to introduce the empirical methods of the sciences into academic philosophy is a *corollary* of moderate scientism.

Likewise, according to the view I call *Weak Scientism* (Mizrahi 2017a), which is the view that scientific knowledge is the *best* (but not the *only*) knowledge we have (Mizrahi 2017a, 354), non-scientific disciplines can and do produce knowledge. However, the best knowledge we have is the knowledge that comes from scientific disciplines. The argument for *Weak Scientism* runs as follows. One thing can be said to be better than another thing either *quantitatively* or *qualitatively*. Accordingly, if scientific knowledge can be said to be better than non-scientific knowledge both quantitatively and qualitatively, then scientific knowledge is better than non-scientific knowledge and *Weak Scientism* is true. Now, scientific knowledge is *quantitatively* better than non-scientific knowledge because scientific disciplines produce *more* knowledge and the knowledge they produce has *more* impact than the knowledge produced by non-scientific disciplines. This claim is supported by data on the research output (i.e., number of publications) and research impact (i.e., citation counts) of academic disciplines. These data show that scientific disciplines produce more publications and those publications get cited more than the publications of non-scientific disciplines (Mizrahi 2017a, 2017b, 2018a, 2018b, 2018c). Indeed, a citation analysis by the *Times Higher Education*'s data team reveals that more than 70 percent of publications in literature and literary theory go uncited after five years, whereas less than 7 percent of publications in electrochemistry go uncited after five years. In general, it is disciplines in the arts and humanities, for the most part, where most publications go uncited after five years, whereas publications in the sciences are almost guaranteed a citation after five years. In philosophy, for example, more than 50 percent of publications go uncited after five years. In behavioral neuroscience, on the other hand, less than 7 percent of publications go uncited after five years. Overall, there are twelve disciplines, mostly in the arts and humanities, in which more than half of publications remain uncited after five years (Baker 2018). Moreover, scientific knowledge is *qualitatively* better than non-scientific knowledge because, unlike non-scientific knowledge, scientific knowledge is explanatorily, predictively, and instrumentally successful (Mizrahi 2017a, 2017b, 2018a, 2018b, 2018c).

On *Weak Scientism*, then, there is no denying that non-scientific disciplines can and do produce knowledge even if they do not use the empirical methods of the sciences. However, if practitioners of non-scientific disciplines want to produce *better* knowledge, then they *better* adopt the empirical methods of the sciences. Again, this is a corollary of *Weak Scientism*. It is not an explicit statement of the view itself, contrary to what Bishop (2019) seems to think. Applied to academic philosophy, then, the call to introduce the empirical methods of the sciences into academic philosophy depends on whether academic philosophers want to produce *better* knowledge than what they have been producing using the traditional methods of philosophy. If they do, then they *better* adopt the empirical methods of the sciences. For, unlike the traditional methods of academic philosophy, the empirical methods of the sciences have a track record of success. Clearly, then, *Weak Scientism*, which is the view that scientific knowledge is the *best* (but not the *only*) knowledge we have and the call for academic philosophers to adopt the empirical methods of the sciences are not one and the same. Nor is the latter an example of the former, as Bishop (2019) seems to think. Instead, the call to introduce the empirical methods of the sciences into academic philosophy is a *corollary* of *Weak Scientism*.

2. METHODOLOGICAL SCIENTISM AND METHODOLOGICAL INTERDISCIPLINARITY

Second, when proponents of scientism call on academic philosophers to adopt the empirical methods of the sciences, they are not merely trying to make academic philosophy more interdisciplinary. For one thing, interdisciplinarity is *not merely* the introduction of the methods of one discipline into another. As one might expect, there are different types of interdisciplinarity. Klein (2017) distinguishes between *methodological interdisciplinarity* and *theoretical interdisciplinarity*. As far as the latter is concerned, Klein (2017) further distinguishes between *instrumental interdisciplinarity* and *critical interdisciplinarity*. Since we are concerned with scientism and the introduction of the empirical methods of the sciences into academic philosophy, our current focus is on methodological interdisciplinarity. According to Klein (2017, 24), methodological interdisciplinarity consists in "*borrowing* a method or concept from another discipline to test a hypothesis, to answer a research question, or to help develop a theory," where the motivation "is to *improve* the quality of results" (emphasis added). Klein (2017, 25) gives the following example:

> New engineering and technological *methods* were also developed during World War II, stimulating postwar *borrowings* of cybernetics, systems theory, information theory, game theory, and new conceptual tools of communication and decision theories. And, the roster of *shared methods* includes techniques such as surveying, interviewing, sampling, polling, case studies, cross-cultural analysis, and ethnography. (emphases added)

For these reasons, philosophy of physics and philosophy of chemistry are *not* clear examples of methodological interdisciplinarity, contrary to what Bishop (2019) seems to think. If philosophers of physics are using the traditional methods of philosophy (e.g., the so-called "method of cases" or appealing to intuitions, the method of case studies, etc.[9]), and they have not borrowed any new methods from other disciplines with the intent of improving the quality of results, then their discipline is *not* interdisciplinary, even if their object of study is physics (as opposed to the typical objects of study for academic philosophers). Similarly, if philosophers of chemistry are using the traditional methods of philosophy (e.g., the method of cases or appealing to intuitions, the method of case studies, etc.), and they have not borrowed any new methods from other disciplines with the intent of improving the quality of results, then their discipline is *not* interdisciplinary, even if their object of study is chemistry (as opposed to the typical objects of study for academic philosophers).

Perhaps a better example of interdisciplinarity in academic philosophy is Philosophy of Science in Practice (PSP). In chapter 6 of this volume, Luana Poliseli and Federica Russo discuss PSP as an example of an interdisciplinary and methodologically diverse field of inquiry within philosophy of science. It should be noted, however, that methodological interdisciplinarity and methodological diversity are not the same thing. A discipline can be methodologically diverse without being methodologically interdisciplinary, and vice versa. For example, PSP may be methodologically diverse insofar as its practitioners employ a variety of different methods, such as the method case studies, conceptual reflection and analysis, and so on, just as Poliseli and Russo argue in chapter 6. But that does not necessarily make PSP methodologically interdisciplinary. For if philosophers of science in practice are simply using the traditional methods of philosophy, such as conceptual analysis and the method of case studies, while not *borrowing* any new methods from other disciplines with the intent of *improving* the quality of results in PSP, then their field is *not* interdisciplinary, even if their object of study is scientific practices rather than theories, which have been the typical object of study for philosophers of science before the so-called "practice turn" in philosophy of science.[10]

Indeed, the empirical methods of the sciences are particularly useful to philosophers of science who are interested in studying scientific practices rather than theories. According to the mission statement of the Society for Philosophy of Science in Practice (2006–2022), "Practice consists of organized or regulated activities aimed at the achievement of certain goals. Therefore, the epistemology of [scientific] practice must elucidate what kinds of activities are required in generating [scientific] knowledge. Traditional debates in epistemology (concerning truth, fact, belief, certainty, observation, explanation, justification, evidence, etc.) may be re-framed with benefit in terms of activities. In a similar vein, practice-based treatments will also shed further light on questions about models, measurement, experimentation, etc., which have arisen with prominence in recent decades from considerations of *actual scientific work*" (emphasis added). Clearly, the empirical methods of the sciences are well suited for studying "actual scientific work" (Philosophy of Science in Practice 2006–2022). For it is difficult to see how one can study actual scientific work and scientific practices by reflecting about them from the armchair. To study actual scientific work and practices, one must observe them. To borrow a phrase from Ludwig Wittgenstein, the philosopher of science in practice must *look and see* what practitioners of a particular scientific discipline actually say and do.

In that respect, perhaps much better examples of interdisciplinarity in PSP may be studies that borrow and use empirical methods from the sciences in order to test a hypothesis or answer a philosophical question about science in an attempt to improve the quality of results pertaining to those hypotheses or questions. For example, in several empirical studies (Mizrahi 2021a, 2022), I have borrowed and used the data mining, data analysis, and data visualization methods of data science in order to test hypotheses about scientific progress. The results of my quantitative, corpus-based studies suggest that, when practicing scientists discuss the aims and goals of their research in their scientific publications, they use the term "understanding" significantly more often than the terms "knowledge," "truth," and "solution." These results lend some empirical support to the so-called noetic account of scientific progress, which defines scientific progress in terms of increasing understanding (Mizrahi 2021a, 2022). Importantly, I have borrowed and used the data mining, data analysis, and data visualization methods of data science to conduct these empirical studies of scientific progress not for the sake of methodological diversity but for the sake of improving the quality of results in philosophy of science. That is, philosophers of science engaged in the debate about scientific progress were using the method of case studies and appealing to intuitions elicited from hypothetical cases, for the most part, when arguing for and/or against philosophical accounts of scientific progress. But it is fair to say that these methods have failed to produce quality results such that most philosophers of science could agree on them. As Prinz (2008, 191) puts it, "In philosophy, debates often collapse into intuition mongering because defenders of opposing views are equally confident about conflicting intuitions." Expectedly, then, philosophers of science have different intuitions elicited by the same "intuition pumps" (or hypothetical cases) and they often use different case studies from the history of science in support of different philosophical accounts of scientific progress.[11] In other words, appealing to intuitions elicited by hypothetical cases and appealing to case studies selected from the history of science have failed "to compel agreement" (Chalmers 2015, 25) as far as the scientific progress debate in philosophy of science is concerned. Accordingly, my borrowing of data mining, data analysis, and data visualization methods from data science was intended to test philosophical accounts of scientific progress in an attempt to improve the quality of results pertaining to those philosophical accounts.

The use of empirical methods in PSP in particular, and in philosophy of science in general, can help to improve the quality of results not only in terms of producing results that philosophers can agree on but also in terms of avoiding preconceived philosophical notions about science. As Rouse (2007, 84) puts it with respect to PSP, "A philosophy of scientific practices [. . .] aims to *avoid unwarranted philosophical impositions upon science*, by attending more closely to what scientists say and do" (emphases added). But the point applies to philosophy of science more generally. In other words, if we *look and see*, rather than reflect about science from the armchair, we will be able to avoid unwarranted philosophical impositions about science. For example, in one experimental study, Wesley Buckwalter and I have borrowed and used the experimental, survey, and statistical methods of psychology to address a philosophical question about the concept of scientific progress, namely, whether the ordinary concept of scientific progress tracks truth alone or truth and justification. The results of our experimental study suggest that people's ordinary conception of scientific progress conforms more closely to the so-called epistemic account of scientific progress, according to which scientific progress is made when scientific knowledge is accumulated, than to the so-called semantic account of scientific progress, according to which the accumulation of truth alone is sufficient for making scientific progress (Mizrahi and Buckwalter 2014). Importantly, we have borrowed the experimental and statistical methods of psychology to conduct this experimental study of scientific progress not for the sake of methodological diversity but for the sake of improving the quality of results in philosophy of science. That is, philosophers of science engaged in the debate about scientific progress were using the method of case studies and appealing to intuitions elicited from hypothetical cases, for the most part, when arguing for and/or against philosophical accounts of scientific progress. But it is fair to say that these methods have not been successful in preventing the imposition of unwarranted philosophical notions upon science. This is because proponents of the so-called epistemic account of scientific knowledge, which defines scientific progress in terms of the accumulation of knowledge, prefer to think of justification along externalist lines, whereas people's intuitive judgments about scientific progress are more sensitive to internal, rather than external, justification (Mizrahi and Buckwalter 2014, 159–160). Accordingly, our borrowing of experimental and statistical methods from psychology was intended to answer a philosophical question about the concept of scientific progress in an attempt to improve the quality of results pertaining to that concept.

For these reasons, the proponents of scientism's call for academic philosophers to borrow the methods of the empirical sciences and employ them in order to test hypotheses and answer research questions in academic philosophy is not to be confused with a mere call for *methodological diversity*. It is not to be confused with a mere call for *methodological pluralism*, either. Some academic philosophers seem to think that academic philosophy is already methodologically pluralistic. For instance, while considering possible future directions philosophy might be headed toward based on his experience as editor of the *European Journal of Philosophy*, Stern (2013, 226) writes that one of the major recent changes has been

> The increased *pluralism* of philosophy, where there is no longer a dominant *area* of philosophy (as philosophy of language was at the high point of the "linguistic turn," for example), or dominant *method* (such as philosophical analysis and phenomenology, which in their day seemed somehow de rigueur), or indeed dominant *ism* or school at all (as opposed to the periods when existentialism or logical positivism or structuralism were in the ascendant). (emphases in original).

On the issue of methodology in particular, Catarina Dutilh Novaes (2012, 255) argues that, in addition to the traditional methods of academic philosophy, namely, conceptual, a priori reflection and analysis (or the so-called method of cases, i.e., appealing to intuitions elicited by hypothetical cases), the following methodological approaches are available to academic philosophers and "should be *combined* in one and the same [philosophical] investigation" (emphases in original):

1. Formal methods: "the application of mathematical and logical tools to the investigation of philosophical issues," for example, formal epistemology (Dutilh Novaes 2012, 251).

2. Historical methods: "tracing the historical origins of a given philosophical concept or practice in order to attain a better understanding of its *current* instantiations," for example, conceptual archeology and genealogy (Dutilh Novaes 2012, 252).
3. Empirically informed methods: "the systematic reference to data and results from the empirical and social sciences," for example, experimental philosophy (Dutilh Novaes 2012, 253).

While Dutilh Novaes (2012) gives experimental philosophy as an example of academic philosophers using empirically informed methods, I would distinguish between *empirically informed philosophy* and *empirically engaged philosophy* (or experimental philosophy). The former is informed by empirical results from the sciences, but it need not *borrow* and *employ* the empirical methods of the sciences, whereas the latter borrows and employs the empirical methods of the sciences in order to test philosophical hypotheses and/or answer philosophical questions.

Insofar as academic philosophers are already making use of formal, historical, and empirically informed methods to some extent, academic philosophy is already methodologically pluralistic. So that cannot be what proponents of scientism are calling for. Instead, what proponents of scientism are calling for is a radical departure from the traditional methods of philosophical inquiry or a reimagining of academic philosophy as no longer an essentially conceptual or a priori field of inquiry. For, if scientific methods are indeed superior to the traditional methods of philosophy, as proponents of scientism argue,[12] then an obvious way to improve the quality of results in academic philosophy is to borrow and use these empirical methods (provided, of course, that academic philosophers do want to improve the quality of results in academic philosophy).

In fact, there is some empirical evidence to suggest that academic philosophers who are using empirically informed methods are producing research that is quantitatively better than the research produced by academic philosophers who are not using empirically informed methods. Recall that one of the reasons to think that scientific knowledge is superior to non-scientific knowledge is that the former is quantitatively better than the latter. More specifically, the research produced in scientific fields is quantitatively better than the research produced in non-scientific fields because scientific fields produce more research than non-scientific fields (as measured by research output or the number of publications) and the research produced in scientific fields has more impact than the research produced in non-scientific fields (as measured by research impact or the number of citations). So, if we take History and Philosophy of Science (HPS) as an example of an area of inquiry within academic philosophy more broadly where practitioners use empirically informed methods (i.e., they are doing empirically informed philosophy, but not necessarily empirically engaged philosophy or experimental philosophy),[13] and we compare it to philosophy in terms of citation counts, we can see that HPS research has more impact than philosophy research. According to SJR (Scimago Journal and Country Rank: https://www.scimagojr.com/), 1,558 documents published in philosophy journals in the last three years (from 2017–2018 to 2019–2020) were cited at least once. In philosophy, the percentage of cited documents in the last three years is 20.85 percent. On the other hand, 14,247 documents published in HPS journals in the last three years (from 2017–2018 to 2019–2020) were cited at least once. In HPS, the percentage of cited documents in the last three years is 42.11 percent. From 2017–2018 to 2019–2020, the average citations per HPS article is 4.76, whereas the average citations per philosophy article is only 0.32. Finally, HPS journals have an average H index of 1,123, whereas philosophy journals have an average H index of only 201. Overall, then, HPS research is quantitatively better than philosophy research in terms of research output (as measured by the number of publications) and research impact (as measured by the number of citations). Clearly, practitioners of HPS are doing something right. Of course, it may not be the use of empirically informed methods that make HPS quantitatively better than philosophy. There may be another explanation for why HPS articles get cited more often than philosophy articles do.

For these reasons, although proponents of scientism are indeed calling on academic philosophers to borrow the empirical methods of the sciences and use these methods to answer philosophical questions and test philosophical hypotheses, to suggest that what they are calling for is merely methodological interdisciplinarity in philosophy, as Bishop (2019) does,[14] is mistaken. Proponents of scientism are indeed calling on academic philosophers to borrow the empirical methods of the sciences and use them in order to test hypotheses and answer philosophical questions with the intent of improving the quality of results in academic philosophy. But they do not stop there. They also argue that the empirical methods of the sciences are superior to the traditional methods of academic philosophy, and so the traditional methods of academic philosophy must give way to the empirical methods of the sciences if academic philosophy is to improve the quality of its results. This is also *not* merely a call for methodological diversity or methodological pluralism. For both methodological diversity and methodological pluralism imply that the traditional methods of academic philosophy and the empirical methods of the sciences can occupy a place of equal methodological importance in academic philosophy. But proponents of scientism would deny that. If some strong version of methodological scientism is true, then the empirical methods of the sciences and the traditional methods of philosophy cannot occupy a place of equal importance, since the former are knowledge-producing methods, whereas the latter are not. And if some weak version of methodological scientism is true, then the empirical methods of the sciences and the traditional methods of philosophy cannot occupy a place of equal importance, since the former are better at producing knowledge than the latter.

3. SCIENTISM AND METHODOLOGICAL DEBATES IN PHILOSOPHY

Finally, the suggestion that proponents of scientism are merely calling for methodological interdisciplinarity, methodological diversity, or methodological pluralism in academic philosophy also betrays a misunderstanding of not only the scientism debate in philosophy but also the methodological debate following the advent of experimental philosophy. As far as the scientism debate is concerned, opponents of scientism are keen to protest what they call "scientific imperialism" (Kitcher 2012) or "scientific expansionism" (Stenmark 2004) and defend the territory of philosophy against any encroachment by the sciences. For example, according to Kitcher (2017, 110–112), scientism makes us underestimate "the impact of the humanities and the arts" and "inspires scientific imperialism." According to Dupré (2001, 16), scientific imperialism is "the tendency for a successful scientific idea to be applied far beyond its original home." In other words, opponents of scientism want science to "stay in its lane," so to speak. Clearly, then, opponents of scientism are unlikely to be on board with calls to borrow and adopt scientific methods in academic philosophy. For them, that would amount to "scientific imperialism" (Kitcher 2012) or "scientific expansionism" (Stenmark 2004), since it would be an instance of science going beyond what is considered to be its proper sphere.[15] But that is precisely what proponents of scientism are calling for. That is, they argue that, if academic philosophy is to improve the quality of its results and produce quality knowledge as the sciences do, then it needs to borrow and employ the empirical methods of the sciences at the expense of its traditional methods of conceptual analysis, appealing to intuitions, and the like.

Indeed, as far as the methodological debate following the advent of experimental philosophy is concerned, some academic philosophers have dug in their heels on this issue precisely. That is, they sought to defend philosophy as an essentially a priori discipline by arguing in defense of the so-called "method of cases," that is, the method of appealing to intuitions elicited by hypothetical cases, in response to the rise of experimental studies that purport to show that intuitions may not be as reliable as philosophers tend to think they are. As Matthew Haug (2014, 1) puts it:

The last few years have seen a surge of interest in philosophical methodology. If there is a single question that unifies the disparate currents of this surge it is whether philosophical questions can successfully be answered "from the armchair."

Academic "philosophers who think that philosophers should remain in their armchairs without substantially engaging with empirical science" have sought to defend "the autonomy and authority of a priori reflection, introspection, or the use of intuition" as evidence for and/or against philosophical hypotheses (Haug 2014, 2). Of course, if philosophy can only be done "from the armchair," by using the traditional methods of philosophy, then it follows that philosophy cannot be done "in the laboratory," by using the empirical methods of the sciences.

Tom Sorell, a prominent opponent of scientism,[16] has made an argument against experimental philosophy along these lines. According to Sorell (2018, 829):

> Is experimental philosophy a kind of philosophy? Since it involves techniques that experimental philosophers themselves admit are not typical of current academic philosophy, and since some of these techniques are borrowed from psychology and other social sciences, it is at least arguable that, despite calling itself "philosophy," experimental philosophy is better classified as psychology or some other social science.[17]

Sorell seems to be arguing that experimental philosophy is not a kind of philosophy because it borrows methods from other disciplines that are different from the methods typically used by academic philosophers. In other words, experimental philosophy is a kind of philosophy only if it uses the methods typically used by academic philosophers. Since experimental philosophers are not using the methods typically used by academic philosophers, it follows that experimental philosophy is not a kind of philosophy. This argument, if sound, ensures that academic philosophy remains an essentially a priori discipline by keeping out the empirical methods of the sciences. Any method that is not typical to academic philosophy is deemed "not philosophy" *ex hypothesi*.[18]

There are several problems with Sorell's argument. More specifically, Sorell's argument seems to be a fallacious appeal to tradition. That is, Sorell seems to be arguing that the traditional methods of academic philosophy best capture what philosophy essentially is. Since the traditional methods of philosophy have been the methods of armchair reflection, conceptual analysis, and the like, doing philosophy essentially means using these methods rather than the empirical methods of the sciences. But why? Why think that academic philosophy must be practiced as it has traditionally been practiced? After all, that is the very question at hand. That is, the question is whether academic philosophy can be practiced not only "from the armchair" but also "in the laboratory." Sorell's premise according to which the practice of philosophy is essentially defined by what academic philosophers have traditionally been doing rules out the possibility of practicing philosophy in the laboratory. Indeed, it rules out introducing any new methods whatsoever into academic philosophy, empirical or otherwise. This means that methodological interdisciplinarity would be ruled out *by definition* if Sorell's argument were sound. Recall that methodological interdisciplinarity consists in "*borrowing* a method or concept from another discipline to test a hypothesis, to answer a research question, or to help develop a theory," where the motivation "is to *improve* the quality of results" (Klein 2017, 24). But Sorell seems to be claiming that a discipline is essentially defined by the methods its practitioners have been using. If one were to use a new method *m* that the practitioners of a particular discipline *D* haven't been using regularly, then one's work would not belong to *D*, by Sorell's lights, even if one has been a practitioner of *D* for many years. If that were true, methodological interdisciplinarity would be impossible. I take this to be an absurd consequence of Sorell's argument.

In fact, the consequences of Sorell's premise, according to which the practice of philosophy is essentially defined by what academic philosophers have traditionally been doing, may be even more absurd than that. This is because Sorell (2018, 829) claims that experimental philosophy is not a kind of philosophy because "it involves techniques that experimental philosophers themselves admit are not typical of *current* academic philosophy" (emphasis added). His use of the term "current" suggests that disciplines cannot undergo methodological changes over time without ceasing to be the discipline they are. To see why, consider radio astronomy. Radio astronomy is a subfield of astronomy in which astronomers study celestial objects that give off radio waves using radio telescopes. Importantly for our purposes, radio astronomy is a relatively young subfield within astronomy. It originated in the early 1930s when Karl G. Jansky, an engineer at Bell Laboratories, noticed noisy static that interfered with short-wave voice communications over the Atlantic Ocean. In his seminal paper, "Radio Waves from Outside the Solar System," Jansky writes that the source of this noise is not the sun, as previously thought (Jansky 1932), but a region outside the solar system (Jansky 1933). Now, to make this groundbreaking discovery, Jansky used techniques or methods that were not typical of astronomy at the time, namely, instruments that developed into what we now call "radio telescopes." By Sorell's reasoning, then, we would have to conclude that whatever Jansky was doing when he made his discovery was not astronomy, since he was using instruments that were not typical of astronomy at the time. In other words, if Sorell's argument shows that experimental philosophy is not a kind of philosophy because experimental philosophers are using techniques that are not typical of current academic philosophy, then it would also show that radio astronomy is not a kind of astronomy because, at the time, Jansky was using techniques that were not typical of contemporary astronomy. I take this to be a refutation by logical analogy of Sorell's argument. In much the same way that radio astronomy is a kind of astronomy, experimental philosophy is a kind of philosophy. As Gonnerman (2018, 465) puts it, "The 'experimental' in 'experimental philosophy' is not like the 'fake' in 'fake diamond'. It's more like the 'good' in 'good ideas'. It modifies the noun to identify the subset." According to proponents of scientism, adopting the empirical methods of the sciences and employing them in academic philosophy is a really good idea.

For an opponent of scientism like Sorell, of course, making academic philosophy more scientific through the adoption of the empirical methods of the sciences is enough to cry "scientism." Indeed, Sorell (2017, 265) accuses experimental philosophers of having a "scientistic tendency." According to Sorell (2017, 265), experimental philosophy is scientistic insofar as "it objectionably treats natural science as the preferred body of results and methods for intellectual work of every kind." However, Sorell's charge against experimental philosophy is misconceived. First, Sorell accuses experimental philosophers of treating "*natural science* as the preferred body of results and methods for intellectual work of every kind" (2017, 265; emphasis added) while, at the same time, also claiming that the "techniques [of experimental philosophy] are borrowed from psychology and other *social sciences*" (2018, 829; emphasis added). But these claims are prima facie inconsistent. On the one hand, if experimental philosophers prefer the methods of the *natural sciences*, why are they borrowing the methods of the *social sciences* and "adopting some elementary experimental techniques from *psychology*"? (Sorell 2017, 265; emphasis added). On the other hand, if experimental philosophers borrow the methods of the *social sciences*, and "experimental techniques from *psychology*" (Sorell 2017, 265; emphasis added), in what sense are they treating "*natural science* as *the preferred* body of results and methods for intellectual work of every kind"? (Sorell 2017, 265; emphasis added).

Second, suppose for the sake of argument that experimental philosophers do treat "natural science as *the preferred* body of results and methods for intellectual work of every kind," as Sorell claims (2017, 265). Why is that objectionable? If there are good reasons to think that scientific methods are superior to the traditional methods of philosophy, then it is not only unobjectionable but also reasonable to adopt scientific methods in academic philosophy. As we have seen, proponents of scientism argue that there are good reasons to believe that scientific methods are better than the traditional methods of philosophy. Briefly, scientific methods are *quantitatively* better than the traditional methods of philosophy because their application generally produces more research (as measured by the number of publications) that has more impact (as measured by the number of citations). Moreover, scientific methods are *qualitatively* better than the traditional methods of philosophy because their application generally produces explanatorily, predictively, and instrumentally successful knowledge.[19] There is nothing objectionable about employing methods with a successful track record of achieving the epistemic goals that both scientists and philosophers are trying to achieve as researchers. After all, as academic fields of inquiry, both science and philosophy are in the business of producing knowledge. Of the traditional methods of philosophy and scientific methods, it is clearly the latter that have a superior track record of epistemic success.

4. CONCLUSION

Proponents of scientism argue that academic philosophy can be done in the laboratory. In fact, doing philosophy in the laboratory would be better for academic philosophy overall because it would improve the quality of its results in two respects. First, borrowing the empirical methods of the sciences and using them to test hypotheses and answer questions in philosophy would improve the quality of results in academic philosophy such that most philosophers could agree on them. Second, borrowing the empirical methods of the sciences and using them to test hypotheses and answer questions in philosophy would improve the quality of results in academic philosophy such that preconceived philosophical notions about the object of study (e.g., science) can be avoided. By helping "to compel agreement" (Chalmers 2015, 25) and avoid preconceived philosophical impositions on its objects of study, perhaps academic philosophy could finally be as successful as science and make progress as science does. Insisting that academic philosophy must be an essentially a priori field of inquiry, and that those who borrow and employ the methods of the empirical sciences are "not doing philosophy," is a territorialist tendency that, if allowed to fester, would only stifle epistemic innovation and hinder progress in academic philosophy.[20]

   As we have seen, the proponents of scientism's call for academic philosophers to adopt the empirical methods of the sciences is not merely a call for methodological interdisciplinarity. Proponents of scientism are not arguing that the empirical methods of the sciences are simply to be added to the toolkit of the academic philosopher. Rather, the empirical methods of the sciences are to radically transform academic philosophy from an essentially a priori discipline to a discipline that looks more like science. Similarly, the proponents of scientism's call for academic philosophers to adopt the empirical methods of the sciences is not merely a call for methodological diversity or methodological pluralism. Proponents of scientism are not arguing that the empirical methods of the sciences are simply to be one method (or set of methods) among many methods used by academic philosophers. Rather, the empirical methods of the sciences are to be the dominant method (or set of methods) academic philosophers use when they do philosophy, since the empirical methods of the sciences are superior to non-scientific methods, either because they are *better* than non-scientific methods, insofar as their use in the sciences produces more impactful knowledge that is explanatorily, predictively, and instrumentally successful (as proponents of weak versions of scientism argue), or because they are the *only* methods that produce knowledge (as proponents of strong versions of scientism argue).

NOTES

1. Transcribed from the following YouTube video: https://youtu.be/sVUMAqMmy7o.

2. For the Aim and Scope of *Metaphilosophy*, see https://onlinelibrary.wiley.com/page/journal/14679973/homepage/productinformation.html.

3. See also the essays collected in Fischer and Collins (2015).

4. For a detailed discussion of the varieties of scientism, see chapter 1 of this volume.

5. See also chapters 3 and 6 of this volume for further discussion.

6. See chapters 1 and 5 of this volume for further discussion.

7. See chapter 1 of this volume for further discussion.

8. See chapter 1 of this volume for further discussion.

9. For more on the so-called method of cases and appeals to intuition in academic philosophy, see Mizrahi (2014), (2015a), (2015b), and (2021b). For more on the method of case studies, see Mizrahi (2020a).

10. See chapter 6 of this volume for further discussion.

11. For more on the method of case studies in philosophy of science, see Mizrahi (2020b, 25–31).

12. See chapter 1 of this volume for a more detailed discussion of the arguments made by proponents of scientism.

13. See chapter 6 of this volume where Luana Poliseli and Federica Russo discuss Philosophy of Science in Practice as an example of empirically informed HPS.

14. See also chapters 3 and 6 of this volume for further discussion.

15. For a more detailed discussion of the "scientific imperialism" charge, see chapter 1 of this volume. Just as opponents of scientism can charge proponents of scientism with "scientific expansionism," proponents of scientism can charge opponents of scientism with "philosophical territorialism." See chapter 2 of this volume.

16. See chapter 1 of this volume for further discussion.

17. See also Sorell (2017). Cf. Sytsma and Livengood (2019).

18. Opponents of scientism and critics of experimental philosophy often complain that empirical work is "not philosophy" (Jenkins 2014) or that it is not "philosophically significant" (Kauppinen 2007). Cf. Knobe (2007) and O'Neill and Machery (2014).

19. See chapter 1 of this volume for a more detailed discussion of the arguments made by proponents of scientism.

20. For more on the "philosophical territorialism" charge, see chapter 2 of this volume.

REFERENCES

Aberdein, Andrew and Matthew Inglis. 2019. "Introduction." In *Advances in Experimental Philosophy of Logic and Mathematics*, edited by Andrew Aberdein and Matthew Inglis, 1–13. London: Bloomsbury.

Baker, Simon. 2018. "How Much Research Goes Completely Uncited?" *Times Higher Education*, April 18, 2018. https://www.timeshighereducation.com/news/how-much-research-goes-completely-uncited.

Bishop, Robert C. 2019. "Scientism or Interdisciplinarity?" *Social Epistemology Review and Reply Collective* 8 (12): 46–49.

Buckwalter, Wesley and John Turri. 2018. "Moderate Scientism in Philosophy." In *Scientism: Prospects and Problems*, edited by J. de Ridder, R. Peels, and R. van Woudenberg, 280–300. New York: Oxford University Press.

Chalmers, David. 2015. "Why Isn't There More Progress in Philosophy?" *Philosophy* 90 (1): 3–31.

Dennett, Daniel C. 2013. *Intuition Pumps and Other Tools for Thinking*. New York: W. W. Norton & Co.

Dolcini, Nevia. 2016. "Philosophy Made Visual: An Experimental Study." In *Model-Based Reasoning in Science and Technology: Logical, Epistemological, and Cognitive Issues*, edited by Lorenzo Magnani and Claudia Casadio, 101–118. Cham, Switzerland: Springer.

Dupré, John. 2001. *Human Nature and the Limits of Science*. Oxford: Oxford University Press.

Dutilh Novaes, Catarina. 2012. *Formal Languages in Logic: A Philosophical and Cognitive Analysis*. New York: Cambridge University Press.

Fischer, Eugene and John Collins. 2015. "Rationalism and Naturalism in the Age of Experimental Philosophy." In *Experimental Philosophy, Rationalism, and Naturalism: Rethinking Philosophical Method*, edited by Eugen Fischer and John Collins, 3–33. New York: Routledge.

Gonnerman, Chad. 2018. "Consciousness and Experimental Philosophy." In *The Routledge Handbook of Consciousness*, edited by Rocco J. Gennaro, 463–475. New York: Routledge.

Haug, Matthew C. 2014. "Debates About Methods: From Linguistic Philosophy to Philosophical Naturalism." In *Philosophical Methodology: The Armchair or the Laboratory?*, edited by Matthew C. Haug, 1–26. New York: Routledge.

Jansky, Karl G. 1932. "Directional Studies of Atmospherics at High Frequencies." *Proceedings of the Institute of Radio Engineers* 20 (12): 1920–1932.

Jansky, Karl G. 1933. "Radio Waves From Outside the Solar System." *Nature* 132 (3323): 66.

Jenkins, Katharine. 2014. "'That's Not Philosophy': Feminism, Academia, and the Double Bind." *Journal of Gender Studies* 23 (3): 262–274.

Kauppinen, Antti. 2007. "The Rise and Fall of Experimental Philosophy." *Philosophical Explorations* 10 (2): 95–118.

Kitcher, Philip. 2012. "The Trouble With Scientism: Why History and the Humanities Are Also a Form of Knowledge." *The New Republic*, May 4, 2012. https://newrepublic.com/article/103086/scientism-humanities-knowledge-theory-everything-arts-science.

Kitcher, Philip. 2017. "The Trouble With Scientism: Why History and the Humanities Are Also a Form of Knowledge." In *Science Unlimited? The Challenges of Scientism*, edited by M. Boudry and M. Pigliucci, 109–120. Chicago: The University of Chicago Press.

Klein, Thompson Julie. 2017. "Typologies of Interdisciplinarity: The Boundary Work of Definition." In *The Oxford Handbook of Interdisciplinarity*, edited by Robert Frodeman, 21–34. New York: Oxford University Press.

Knobe, Joshua. 2007. "Experimental Philosophy and Philosophical Significance." *Philosophical Explorations* 10 (2): 119–121.

Mizrahi, Moti. 2014. "Does the Method of Cases Rest on a Mistake?" *Review of Philosophy and Psychology* 5 (2): 183–197.

Mizrahi, Moti. 2015a. "Don't Believe the Hype: Why Should Philosophical Theories Yield to Intuitions?" *Teorema: International Journal of Philosophy* 34 (3): 141–158.

Mizrahi, Moti. 2015b. "Three Arguments Against the Expertise Defense." *Metaphilosophy* 46 (1): 52–64.

Mizrahi, Moti. 2017a. "What's So Bad About Scientism?" *Social Epistemology* 31 (4): 351–367.

Mizrahi, Moti. 2017b. "In Defense of Weak Scientism: A Reply to Brown." *Social Epistemology Review and Reply Collective* 6 (2): 9–22.

Mizrahi, Moti. 2018a. "More in Defense of Weak Scientism: Another Reply to Brown." *Social Epistemology Review and Reply Collective* 7 (4): 7–25.

Mizrahi, Moti. 2018b. "Weak Scientism Defended Once More." *Social Epistemology Review and Reply Collective* 7 (6): 41–50.

Mizrahi, Moti. 2018c. "Why Scientific Knowledge is Still the Best." *Social Epistemology Review and Reply Collective* 7 (9): 18–32.

Mizrahi, Moti. 2019. "What Isn't Obvious about 'Obvious': A Data-Driven Approach to Philosophy of Logic." In *Advances in Experimental Philosophy of Logic and Mathematics*, edited by A. Aberdein and M. Inglis, 201–224. London: Bloomsbury.

Mizrahi, Moti. 2020a. "The Case Study Method in Philosophy of Science: An Empirical Study." *Perspectives on Science* 28 (1): 63–88.

Mizrahi, Moti. 2021a. "Conceptions of Scientific Progress in Scientific Practice: An Empirical Study." *Synthese* 199 (1–2): 2375–2394.

Mizrahi, Moti. 2020b. *The Relativity of Theory: Key Positions and Arguments in the Contemporary Scientific Realism/Antirealism Debate*. Cham, Switzerland: Springer.

Mizrahi, Moti. 2021b. "Your Appeals to Intuition Have No Power Here!" *Axiomathes*. DOI: 10.1007/s10516-021-09560-9.

Mizrahi, Moti. 2022. "What is the Basic Unit of Scientific Progress? A Quantitative, Corpus-Based Study." *Journal for General Philosophy of Science*. DOI: https://doi.org/10.1007/s10838-021-09576-0.

Mizrahi, Moti and Wesley Buckwalter. 2014. "The Role of Justification in the Ordinary Concept of Scientific Progress." *Journal for General Philosophy of Science* 45 (1): 151–166.

O'Neill, Elizabeth and Edouard Machery. 2014. "Experimental Philosophy: What is it Good For?" In *Current Controversies in Experimental Philosophy*, edited by Elizabeth O'Neill and Edouard Machery, vii–xxix. New York: Routledge.

Prinz, Jesse J. 2008. "Empirical Philosophy and Experimental Philosophy." In *Experimental Philosophy*, edited by Joshua Knobe and Shaun Nichols, 189–208. New York: Oxford University Press.

Rouse, Joseph. 2007. "Naturalism and Scientific Practices: A Concluding Scientific Postscript." In *Naturalized Epistemology and Philosophy of Science*, edited by C. M. Mi and Ruey-lin Chen, 61–86. Amsterdam: Rodopi.

Society for Philosophy of Science in Practice. 2006–2022. "Mission Statement." *Society for Philosophy of Science in Practice*. Accessed January 31, 2022. https://philosophy-science-practice.org/about/mission-statement.

Sorell, Tom. 2017. "Scientism (and Other Problems) in Experimental Philosophy." In *Science Unlimited? The Challenges of Scientism*, edited by M. Boudry and M. Pigliucci, 263–282. Chicago: The University of Chicago Press.

Sorell, Tom. 2018. "Experimental Philosophy and the History of Philosophy." *British Journal for the History of Philosophy* 26 (5): 829–849.

Stenmark, Mikael. 2004. *How to Relate Science and Religion: A Multidimensional Model*. Grand Rapids, MI: Wm. B. Eerdmans Publishing Co.

Stern, Robert. 2013. "Whither Philosophy?" *Metaphilosophy* 44 (3): 222–229.

Sytsma, Justin and Jonathan Livengood. 2019. "On Experimental Philosophy and the History of Philosophy: A Reply to Sorell." *British Journal for the History of Philosophy* 27 (3): 635–647.

Van De Poel, Ibo. 2020. "Should Philosophers Begin to Employ New Methods If They Want to Become More Societally Relevant?" In *Philosophy in the Age of Science? Inquiries into Philosophical Progress, Method, and Societal Relevance*, edited by Julia Hermann, Jeroen Hopster, Wouter Kalf, and Michael Klenk, 231–244. London: Rowman & Littlefield.

Chapter 10

Epilogue

Moti Mizrahi

In a contribution to a special issue of *Metaphilosophy* on "The Future of Philosophy: Metaphilosophical Directions for the Twenty-First Century," Philip Kitcher argues that "our current image of philosophy should be turned inside out" (2011, 259). Focusing on Anglophone academic philosophy in the analytic tradition in particular, Kitcher (2011, 253) argues that the central questions of academic philosophy, that is, abstract questions about knowledge, language, and mind, should give up their central place to "questions that are urgent for all people—or at least for all people who have any chance of directing the course of their lives." More specifically, instead of abstract questions like "What is knowledge?" which Kitcher (2011, 255–256) deems *individual*, academic philosophers should reorient themselves toward questions that are *social* like "How are the claims of expertise to be balanced against the claims of democracy?" These social questions, Kitcher (2011, 258) argues, are the important questions "that philosophy should be addressing, and that much of what is taken to lie at the center of our subject has no obvious bearing on any such question." If academic philosophers were to turn academic philosophy inside out in this way, Kitcher (2011, 259) argues, the result would be an academic philosophy with some social relevance.[1]

Interestingly, however, Kitcher does not think that transforming academic philosophy, so as to make it more societally relevant, requires any attention to or change in methodology. As Kithcer (2011, 259) puts it, "What binds these [philosophical] endeavors [that have social relevance] together is a concern for philosophical *questions* that matter, rather than a shared *method*" (emphases added). This is peculiar because Kitcher (2011, 251) himself admits that scientists "see themselves as having 'methods' for arriving at reliable results," whereas "Philosophy isn't like that." But as long as academic philosophy does not have methods for arriving at reliable results, as Kitcher seems to admit, it is difficult to see how it can provide satisfactory answers to questions that matter to people. In other words, if academic philosophy is to provide satisfactory answers to "questions that are particularly salient for people," as Kitcher (2011, 255) wants it to, then academic philosophy needs methods for arriving at reliable results—or at least methods that academic philosophers can accept as providing reliable results; otherwise, it is difficult to see how academic philosophy can provide satisfactory answers to "questions that are urgent *for all people*" (Kitcher 2011, 253; emphases added).

Ibo Van De Poel (2020) agrees. Van De Poel argues that a methodological renewal is needed if academic philosophy is to become more societally relevant. One reason to think that academic philosophy must undergo a methodological transformation if it is to become more societally relevant has to do with a method widely used by academic philosophers. That method is the so-called "method of cases," that is, appealing to intuitions elicited from hypothetical cases. In addition to many other problems with the so-called method of cases,[2] Van De Poel (2020, 232) adds the following: the hypothetical cases that academic philosophers tend to imagine are far removed from real-world cases. Such hypothetical cases, then, are not a good guide for answering questions that matter to people. In other words, academic philosophers tend to consider what Van De Poel (2020, 233) calls "toy problems," which are different from real-world problems. If they want to solve real-world problems, academic philosophers must abandon their imagined "toy examples" and "toy problems"; instead, academic philosophers must consider real-world problems. Doing so, Van De Poel (2020) argues, requires that academic philosophers adopt methods that involve design and experimentation. By "design," Van De Poel (2020, 243) means "planned and deliberated activities aimed at changing the world in a particular way that is deemed desirable." By "experimentation," Van De Poel (2020, 240) means "trying out something (e.g. a course of action), observing the consequences, and learning from it." Both design and experimentation, Van De Poel (2020, 241) argues, "have a proper place in philosophical inquiries that are aimed at addressing real-world problems."

The introduction of design and experimentation techniques in particular, as well as empirical methods in general, into academic philosophy need not pose an existential threat to academic philosophy.[3] Unless one thinks that academic philosophy is essentially a conceptual or a priori area of inquiry, and it must remain so come what may, there is no immediate existential threat to academic philosophy posed by the adoption of the empirical methods of the sciences by academic philosophers. For this reason, it is difficult to see accusations of "that is not philosophy" or "that is not philosophically significant" leveled against academic philosophers who use empirical methods in their research as any more than a misguided tendency for territorialism. For example, experimental philosophers who use experimentation and statistical methods from the social sciences, such as psychology, in their philosophical work are often called by some academic philosophers (see, e.g., Kauppinen 2007) to explain and justify the philosophical significance of their work (see, e.g., Knobe 2007). Likewise, empirical philosophers who use corpus-based and digital methods from data science and computational linguistics in their philosophical work (see, e.g., Mizrahi 2021) also face "that is not philosophy" or "that is not philosophically significant" calls from academic philosophers who think that empirical methods have no bearing on philosophical questions (see, e.g., Niiniluoto 2019).[4]

Practitioners of an academic discipline that has no methods for arriving at reliable results, however, are not in a position to make such accusations. If academic philosophy had established methods of inquiry that academic philosophers agreed on, then perhaps the introduction and adoption of new methods from other disciplines would not be warranted. But, as Kitcher (2011, 251) puts it, "Philosophy isn't like that." To put the point in the terminology of History and Philosophy of Science (HPS), which stems from Thomas Kuhn's *The Structure of Scientific Revolutions* (1962/1996), academic philosophy is an immature discipline rather than a mature discipline. Mature academic disciplines are "based upon conceptually integrated paradigms, commonly accepted research practices, standardized problem definitions, canonized exemplary solutions, and binding types of theoretical explanations," whereas "the more diffuse, personalized, and idiosyncratic are the standards of epistemic legitimacy," the more likely an academic discipline is to be viewed as immature (Fuchs and Turner 1986, 148). Insofar as academic philosophy does not have "conceptually integrated paradigms, commonly accepted research practices, standardized problem definitions, canonized exemplary solutions, and binding types of theoretical explanations" (Fuchs and Turner 1986, 148), it cannot be considered a mature academic discipline. Likewise, insofar as academic philosophy's standards of epistemic legitimacy depend in large part on idiosyncratic toy examples, toy problems, and intuitive judgments about such toy examples and toy problems, it cannot be considered a mature academic discipline.

In the absence of commonly accepted research practices and methods, then, the introduction of new methods into academic philosophy is not only warranted but also reasonable, especially methods with a track record of success, such as the successful track record of the empirical methods of the sciences. Indeed, there is already evidence to suggest that fields within academic philosophy, where academic philosophers use empirically informed (as in Philosophy of Science in Practice, for example[5]) and/or empirically engaged methods (as in experimental philosophy, for example[6]), are seeing improved results. Accordingly, if academic philosophy is to become a more mature academic discipline, as well as more societally relevant (Van De Poel 2020), it better adopt the empirical and experimental methods of the sciences. The adoption of the empirical and experimental methods of the sciences by academic philosophers need not "spell shipwreck for philosophy itself" (Haack 2017, 43), as some fear. Academic philosophers can be trained to do empirically informed, empirically engaged, and societally relevant work, such that arbitrary disciplinary boundaries can be done away with, and academic philosophy and empirical science become complementary.[7] That is what some proponents of scientism are calling for. That is, if scientific methods are indeed superior to the traditional methods of philosophy, as proponents of scientism argue,[8] then an obvious way to improve the quality of results in academic philosophy is to borrow and use these empirical methods (provided, of course, that academic philosophers do want to improve the quality of results in academic philosophy). For this to happen, however, there must be no room for philosophical territorialism in academic philosophy. For academic philosophy to succeed, disciplinary border policing must be abandoned.[9]

NOTES

1. See chapter 7 of this volume for further discussion on societal problems for science and the humanities.

2. See chapter 9 of this volume for further discussion on the so-called "method of cases."

3. See chapter 3 of this volume for further discussion on the threat of so-called "scientism expansionism."

4. See chapter 2 of this volume for further discussion of the use of corpus-based methods in philosophy.

5. See chapter 6 of this volume for further discussion on empirically informed HPS.

6. See chapter 9 of this volume for further discussion on empirically engaged philosophy.

7. See chapter 8 of this volume for further discussion on cooperation between science and the humanities.

8. See chapters 1, 5, and 9 of this volume for a more detailed discussion of the arguments made by proponents of scientism.

9. See chapter 4 of this volume for further discussion on various conceptions of philosophy.

REFERENCES

Fuchs, Stephan and Jonathan H. Turner. 1986. "What Makes a Science 'Mature'?: Patterns of Organizational Control in Scientific Production." *Sociological Theory* 4 (2): 143–150.

Haack, Susan. 2017. "The Real Question: Can Philosophy Be Saved?" *Free Inquiry* 37 (6): 40–43.

Kauppinen, Antti. 2007. "The Rise and Fall of Experimental Philosophy." *Philosophical Explorations* 10 (2): 95–118.

Kitcher, Philip. 2011. "Philosophy Inside Out." *Metaphilosophy* 42 (3): 248–260.

Knobe, Joshua. 2007. "Experimental Philosophy and Philosophical Significance." *Philosophical Explorations* 10 (2): 119–121.

Kuhn, Thomas, S. 1962/1996. *The Structure of Scientific Revolutions*. Third Edition. Chicago: The University of Chicago Press.

Mizrahi, Moti. 2021. "Conceptions of Scientific Progress in Scientific Practice: An Empirical Study." *Synthese* 199 (1–2): 2375–2394.

Niiniluoto, Ilkka. 2019. "Scientific Progress." In *The Stanford Encyclopedia of Philosophy*, edited by Edward N. Zalta. Winter 2019 Edition. https://plato.stanford .edu/archives/win2019/entries/scientific-progress/.

Van De Poel, Ibo. 2020. "Should Philosophers Begin to Employ New Methods If They Want to Become More Societally Relevant?" In *Philosophy in the Age of Science? Inquiries into Philosophical Progress, Method, and Societal Relevance*, edited by Julia Hermann, Jeroen Hopster, Wouter Kalf, and Michael Klenk, 231–244. London: Rowman & Littlefield.

Index

About the Editor

Moti Mizrahi is associate professor of philosophy at the Florida Institute of Technology. He is the author of *The Relativity of Theory: Key Positions and Arguments in the Contemporary Scientific Realism/Antirealism Debate* (Springer, 2020) and the editor of *The Kuhnian Image of Science: Time for a Decisive Transformation?* (Rowman & Littlefield, 2018). He has published extensively on the philosophy of science, the scientific realism/anti-realism debate, the epistemology of philosophy, and argumentation. His work has appeared in journals such as *Argumentation, British Journal for the Philosophy of Science, Erkenntnis, Philosophical Studies, Studies in History and Philosophy of Science*, and *Synthese*. A major theme in his work has been the application of digital, statistical, and data-driven methods to problems concerning moral, philosophical, and scientific reasoning.

About the Contributors

Ann-Sophie Barwich is a cognitive scientist and empirical philosopher and historian of science, technology, and the senses. Her research centers on the conceptual foundations of neuroscience and theories of perception. In particular, she explores how our understanding of mind and brain would be different if we were looking at the sense of smell. Before starting as an assistant professor at IUB, Barwich was a presidential scholar in Society and Neuroscience at The Center for Science & Society, Columbia University, and held a Research Fellowship at the Konrad Lorenz Institute for Evolution and Cognition Research, Vienna.

Amanda Bryant works primarily on the epistemology and methodology of metaphysics, as well as on naturalisms and their relations to nearby stances such as empiricism, physicalism and scientism. She completed her PhD at the CUNY Graduate Center. Since then, she has held a teaching position at Trent University and a postdoctoral fellowship at the University of Lisbon. She now teaches at Ryerson University. Some of her notable publications include "Keep the Chickens Cooped: The Epistemic Inadequacy of Free Range Metaphysics" (2020) in *Synthese* and "Epistemic Infrastructure for a Scientific Metaphysics" (forthcoming) in *Grazer Philosophische Studien*.

Johan Hietanen is a graduate of the University of Helsinki. Besides philosophy, he has also studied physics. He specializes in philosophy of science and quantum mechanics, which are also topics of his ongoing doctoral dissertation project. Hietanen is the acting president of the Helsinki Circle.

Ilmari Hirvonen is a doctoral student at the University of Helsinki. Besides philosophy, he has some background in general linguistics. His main interests lie in philosophy of pseudoscience and epistemic justification, but he has also worked on metaphilosophy, history of empiricism, philosophy of language, and philosophy of religion. Hirvonen is the acting secretary of the Helsinki Circle.

Ian James Kidd is a lecturer in philosophy at the University of Nottingham, having previously taught at Durham and Leeds. His research interests include topics in epistemology, philosophy of science, and Continental European philosophy. Recent edited collections include *Wittgenstein and Scientism* (with Jon Beale) and *Science and the Self: Animals, Evolution, and Ethics: Essays in Honour of Mary Midgley* (with Liz McKinnell).

Ilkka Pättiniemi studies philosophy at the University of Helsinki. Besides philosophy, he has studied theoretical physics and mathematics. His areas of interest include metaphilosophy, justification, philosophy of science, the foundations of quantum mechanics, and Richard Rorty. He has coauthored papers on the philosophy of physics and on structural realism. Pättiniemi is a member of the Helsinki Circle.

Luana Poliseli is a postdoctoral researcher at the Konrad Lorenz Institute for Evolution and Cognition Research (KLI), Austria, working in the project Ecological understanding as key to improve sustainability sciences. Her interests lie in approaching general philosophical questions through empirical knowledge of particular sciences, focusing on aesthetics and visualization in scientific understanding and model-building for sustainability sciences. She is also a member of the GEOS—Global Epistemologies and Ontologies of Science, at Wageningen University & Research, Netherlands; and part of the National Institute of Science and Technology in Inter- and Transdisciplinary Studies in Ecology and Evolution (INCT-INTREE), Brazil. There she engages with methodological aspects of inter- and transdisciplinary research and with the epistemology of Traditional Ecological Knowledge through an account of Traditional Ecological Understanding.

Federica Russo is a philosopher of science and technology based at the University of Amsterdam. She has a long-standing interest in social science and medical methodology, and she wrote extensively about causal modelling, explanation, and evidence in the social, biomedical, and policy sciences. Among her contributions are *Causality and Causal Modelling in the Social Sciences. Measuring Variations* (Springer, 2009), *Causality: Philosophical Theory Meets Scientific Practice* (OUP, 2014, with P. Illari), and *Evaluating*

Evidence of Mechanisms in Medicine: Principles and Procedures (Springer, 2018, a coauthored monograph of the EBM+ group). Together with P. Illari, she is editor-in-chief of the *European Journal for Philosophy of Science*, and she is an executive editor of *Philosophy and Technology*.

Henrik Saarinen has a master's degree in philosophy from the University of Helsinki. Besides philosophy, he has studied cognitive science and statistics. His areas of interest include general philosophy of science, metametaphysics, and metaphilosophy. Saarinen is a member of the Helsinki Circle.

Petri Turunen is a doctor of theoretical philosophy from the University of Helsinki. He also has a background in theoretical physics (MSc), mathematics, and in some theoretical ecology. His areas of interest include general philosophy of science, philosophy of physics and mathematics, interdisciplinarity, analytic methodology, and metaphilosophy. Turunen is a member of the Helsinki Circle.

Catherine Wilson is presidential professor at the Graduate Center of the City University of New York (CUNY). Her research is focused on the relationship between the natural and social sciences and classical philosophical problems in the following areas: the history of atomism and materialism; theories of life and sensory awareness; visuality and aesthetics, and morality and the emotions from an evolutionary perspective. She has published numerous books and articles on these topics. Her most recent book is *How to Be an Epicurean* (Basic Books, 2019).